Botany - Recent Advances and Applications

Edited by Bimal Kumar Ghimire

Published in London, United Kingdom

IntechOpen

Supporting open minds since 2005

Botany - Recent Advances and Applications
http://dx.doi.org/10.5772/intechopen.87519
Edited by Bimal Kumar Ghimire

Contributors
Jyoti Rani, Sinan Meriç, Alp Ayan, Tamer Gümüş, Çimen Atak, Bezaye Tessema, Brian Wilson, Heiko Daniel, Zenebe Adimassu, El Hadji Malick CISSE, Fan Yang, Da-Dong Li, Ling-Feng Miao, Inam Ullah Khan, Syed Aftab Hussain Shah

First published in London, United Kingdom, 2021 by IntechOpen
IntechOpen is the global imprint of INTECHOPEN LIMITED, registered in England and Wales, registration number: 11086078, 5 Princes Gate Court, London, SW7 2QJ, United Kingdom
Printed in Croatia

British Library Cataloguing-in-Publication Data
A catalogue record for this book is available from the British Library

Additional hard and PDF copies can be obtained from orders@intechopen.com

Botany - Recent Advances and Applications
Edited by Bimal Kumar Ghimire
p. cm.
Print ISBN 978-1-83969-258-1
Online ISBN 978-1-83969-259-8
eBook (PDF) ISBN 978-1-83969-260-4

For EU product safety concerns:
IN TECH d.o.o., Prolaz Marije Krucifikse Kozulić 3, 51000 Rijeka, Croatia,
info@intechopen.com or visit our website at intechopen.com.

Meet the editor

Dr. Bimal Kumar Ghimire is currently an assistant professor in the Department of Crop Science, College of Sanghuh Life Science, Konkuk University, Seoul, South Korea. He received his MS in Botany (major in Cytogenetics) from North Bengal University, India. He obtained his Ph.D. in Agriculture Science from Kangwon National University, South Korea, in 2008. Subsequently, he joined the Department of Botany, Sikkim University, India, as an assistant professor. He has published more than 100 articles in peer-reviewed journals and books. He has edited and written several books and contributed chapters on different aspects of plant sciences including light sources generally used in controlled agriculture, genetic transformation, and secondary metabolites of medicinal plants. His research interests and experience include plant anatomy, genetic diversity in bioenergy crops, allyl compounds, tissues and cell culture, bioreactor culture, and genetic transformation of important medicinal plants.

Contents

Preface

Recent advances in the field of plant biotechnology succeeded in altering various metabolic pathways via gene and metabolic engineering approaches to introduce novel compound production in plants. Molecular biotechnology strategies can be used to improve plant biomass by enhancing cellulose and starch for biofuel production. Biotechnology and genetic engineering approaches are useful for the improvement of crops for biotic and abiotic stress tolerance and increased food production. As such, this book provides updates on recent developments in plant science highlighting major trends in botany, particularly in plant biotechnology, alternative energy, food industry, environment, and medicine. *Botany - Recent Advances and Applications* presents comprehensive information about current advances in plant science. It contains five chapters by eminent research scholars that cover a vast range of research on soil organic carbon sequestration of tropical grasses and their potential use as biomass energy crops to mitigate global environmental changes. The book also highlights nonedible seeds as a potential renewable source of biodiesel production to control the global energy crisis. Finally, the book discusses the importance of microRNA as a potential tool for improving agronomic traits of several economically important crop plants as well as examines the uses, methods, and advantages of recombinant DNA technology and plant biotechnology for the development of plant-based vaccines.

We are enormously grateful to our publisher IntechOpen and the authors who contributed their work to this book. We also extend our sincerest appreciation to Editor and Author Service Manager Sara Gojevic-Zrnic, who was instrumental in the preparation and publication of this volume.

Dr. Bimal Kumar Ghimire
Department of Crop Science,
College of Sanghuh Life Science,
Konkuk University,
Seoul, Korea

Chapter 1

Insights into Metabolic Engineering of the Biosynthesis of Glycine Betaine and Melatonin to Improve Plant Abiotic Stress Tolerance

Cisse El Hadji Malick, Miao Ling-Feng, Li Da-Dong and Yang Fan

Abstract

Metabolic engineering in plant can be describe as a tool using molecular biological technologies which promotes enzymatic reactions that can enhance the biosynthesis of existing compounds such as glycine betaine (GB) in plant species that are able to accumulate GB, or produce news compounds like GB in non-accumulators plants. Moreover we can include to these definition, the mediation in the degradation of diverse compounds in plant organism. For decades, one of the most popular ideas in metabolic engineering literature is the idea that the improvement of gly betaine or melatonin accumulation in plant under environmental stress can be the main window to ameliorate stress tolerance in diverse plant species. A challenging problem in this domain is the integration of different molecular technologies like transgenesis, enzyme kinetics, promoter analysis, biochemistry and genetics, protein sorting, cloning or comparative physiology to reach that objective. A large number of approaches have been developed over the last few decades in metabolic engineering to overcome this problem. Therefore, we examine some previous work and propose some understanding about the use of metabolic engineering in plant stress tolerance. Moreover, this chapter will focus on melatonin (Hormone) and gly betaine (Osmolyte) biosynthesis pathways in engineering stress resistance.

Keywords: metabolic engineering, biosynthesis, molecular, abiotic stress, stress tolerance

1. Introduction

The global climate change influence negatively plant growth and development via the increase of the intensity of various abiotic stresses such as drought, chilling, salinity, waterlogging or flooding. Environmental stresses are one of the most threatening factors that can cause massive losses in agricultural crop production, ranging from 50–70% [1]. Plant biotechnology and engineering are promising platform for exploring the unlimited potential of many various plants species [2]. In recent years, plant metabolism engineering provides successful pathways to increase the production of metabolites that can significantly counterattack the damages caused by diverse abiotic stresses [3]. To improve stress tolerance in plant, various metabolic engineering technologies were used to introduce or increase the synthesis of diverse osmolytes, secondary metabolites or hormones. The adaptation of various plant species to stressful environments can be managed through: (i) the identification of diverse mechanisms developed by plants to counterbalance abiotic stresses (ii) and the improvement of these processes in plants by metabolic engineering [4, 5]. Plant by-products including hormone (melatonin, MT) and osmoprotectant (glycine betaine, GB) that play a prominent roles in plant stress tolerance have been targeting in various plant species to counterattack environmental stresses. The clarification of the biosynthetic pathway of various plant compounds has provided the possibility to metabolically engineer new capabilities in plants as well as successfully engineer whole pathways into microbial systems [6]. Under environmental stresses plant is able to accumulate different molecules such as melatonin or glycine betaine to provide stress tolerance by counteracting with oxidative stress caused by drought, chilling, salinity or heavy metal stresses [7–9]. The protective properties of GB and MT in plant under abiotic stresses had made these substances targets for plant engineering resistance.

The natural biosynthesis of glycine betaine takes place in marine algae and various higher plant species belong to diverse families, counting the Gramineae, Malvaceae, Asteraceae, or Amaranthaceae [10–14]. Glycine betaine accumulation in non-accumulators and accumulators plant species under environmental stresses has long been a target for engineering stress resistance [15, 16]. The biosynthesis of glycine betaine passes by choline → betaine aldehyde → glycine betaine pathways. Most of the enzymes involving in these pathways such as choline monooxygenase (CMO) or betaine aldehyde dehydrogenase (BADH) have been identified, and genes for some of them have been cloned [4, 13].

Indeed, GB as a non-toxic molecule is biosynthesized through two phases of choline oxidation: the first step (Choline → betaine aldehyde) is catalyzed by CMO, and the second step (Betaine aldehyde → glycine betaine) is activated by BADH [13, 17]. The expression of CMO or BADH in tobacco has been done via the cDNA from two natural glycine betaine accumulators; spinach and sugar beet plants. The 35S promoter from plant virus, cauliflower mosaic virus which is a fundamental element of transgenic constructs in the majority of genetically modified plant species was used in transgenic tobacco to control the expression of cDNA for BADH pathway [18]. Also, a crucial tool in metabolism engineering of glycine betaine pathway is the use of a single gene codA from *Arthrobacter globiformis* which is involved in the synthesis of GB [19]. However, GB accumulation in transgenic plants depends on the capacity of endogenous choline uptake, the type of gene that catalyzes the GB biosynthetic pathway, and the localization of the transgene product in a particular cellular compartment [20].

Melatonin a plant hormone identified in a wide variety of animals and plants, has been extensively studied in plants for its properties to counteract with various environmental and biotic stresses [21, 22]. Transcriptome analyses indicated that melatonin primarily affects the pathways of plant hormone signal transduction and biosynthesis of secondary metabolites [23]. In plant the biosynthesis of melatonin is initiated with tryptophan which is converted in serotonin, and between the tryptophan and melatonin, the enzymes hydroxyindole-*O*-methyltransferase and caffeic acid O-methyltransferase (ASMT/COMT) catalyzed a reaction that produce an intermediate molecule named 5-methoxytryptamine [24–26]. The related enzymes involved in melatonin biosynthesis pathway have been targeted to improve stress tolerance in diverse plant species. The over expression of COMT like gene (TaCOMT) in a transgenic *Arabidopsis* via various metabolic engineering techniques (cloning, transgenesis, genetics or promoter analysis) provided drought tolerance by increasing the concentration of melatonin [27]. Other enzymes such as serotonin N-acetyltransferase (MsSNAT) involve in melatonin biosynthesis have been targeted in rice [28] or *Arabidopsis* [29] to provide stress tolerance, either to clarify the role of melatonin in plant. This chapter will focus on the use of glycine betaine, spermidine and melatonin in plant metabolism engineering, particularly in stress engineering.

2. Glycine betaine and metabolism engineering

Glycinebetaine is a quaternary ammonium compound that appears commonly in a large diversity of plants, animals and microorganisms, the first betaine discovered was trimethylglycine (**Figure 1**) named also N, N,N-trimethylglycine [8, 12]. The glycine betaine as a osmolytes is a crucial non-toxic molecule that is accumulated in various plant species under environmental stresses [15].

2.1 Glycine betaine biosynthesis

GB synthesis begins with an essential molecule named choline, synthesized through three sequential adenosyl-methionine dependent methylations of phospho-ethanolamine catalyzed by the cytosolic enzyme phosphoethanolamine methyltransferase (phosphoethanolamine N-methyltransferase) [30]. In plant, the biosynthesis of GB is two steps of oxidation initiated with choline and then betaine aldehyde (**Figure 2**). In plant such as *Arabidopsis* the biosynthesis of choline can be resume by this following line: L-serine → ethanolamine → *O*-phosphoethanolamine → N-methylethanolamine phosphate → N-dimethylethanolamine phosphate → phosphocholine → choline [31–33]. Pursuing the transformation of N-methylethanolamine phosphate by phosphoethanolamine methyltransferase (PeMt) the byproduct differs according to the plant species, for instance in that stage the spinach produce choline like in *Arabidopsis* choline biosynthesis pathway, meanwhile in tobacco PeMt catalyzed a reaction that synthesize phosphatidylcholine in the first place then metabolized to choline [8, 15]. The first stage of GB biosynthesis is modulated by CMO which is an Fd-dependent monooxygenase with a Rieske-type iron–sulfur (2Fe-2S) cluster-binding motif. The second stage of GB

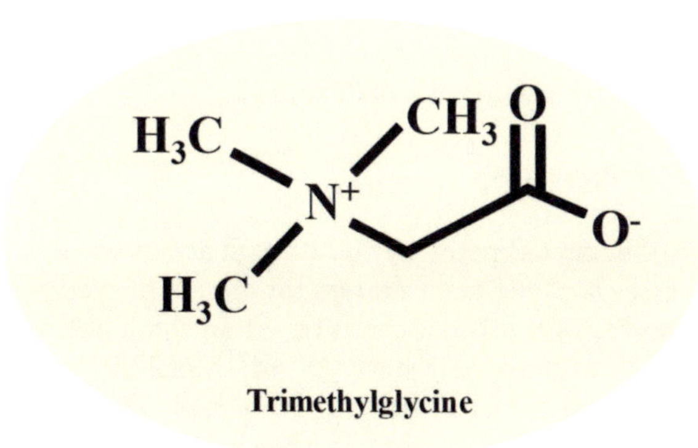

Figure 1.
N,N,N-trimethylglycine.

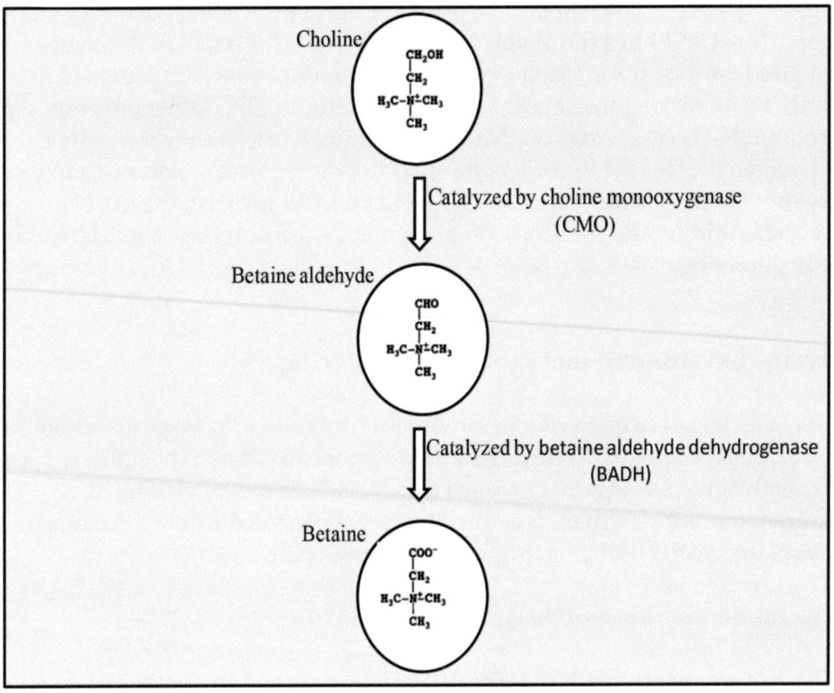

Figure 2.
Diagram of GB biosynthesis in brief.

biosynthesis is catalyzed by BADH, an enzyme belong to the superfamily of aldehyde dehydrogenases which is an NAD^+ or $NADP^+$ dependent [17, 34].

2.2 Glycine betaine and environmental stress

Many plants are able to accumulate naturally GB and diverse osmoprotectants to balance the disruption of plant cell homeostasis caused by environmental stress such as drought, chilling, salinity or high temperature [8, 35, 36]. Many studies have been reported on the positive effect of endogenous GB in plants under abiotic stresses. The role of glycine betaine in osmotic adjustment was related in *Amaranthus tricolor* [37] and *Hordeum maritimum* [38] under salinity. The role of GB against oxidative stress via scavenging the reactive oxygen species and increasing the antioxidant activities was reported in many studies [39–41]. For these reasons, the use of glycine betaine in non accumulator and accumulator plant species become more popular in plant physiology. Indeed, several reports have related the positive effect of GB in transgenic plants (**Table 1**).

2.3 Glycine betaine engineering

The idea of introducing GB pathway and its high accumulation in plant under environmental stresses has long been a target for metabolism engineering stress tolerance. The feasibility of this process was based on comparative physiology and genetic evidence from a maize mutant [15, 54]. Metabolic engineering of the biosynthesis of GB from choline by using various genes such as cod*A* or BADH gene gained more attention to improve stress tolerance in crop and woody plants that are incapable of synthesizing GB under abiotic stresses [8, 18, 55]. Moreover, genetic engineering is also use to increase GB accumulation in various plant species which

Transgenic species	GB Acc./ GB N-Acc.	Type of abiotic stress	Role in stress tolerance	References
Nicotiana tabacum	GB N-Acc.	Salinity	Protection of the photosynthetic apparatus	[42]
Zea mays	GB Acc.	Chilling stress	Protect photosynthesis, Homeostasis	[43]
Synechococcus sp.	GB N-Acc.	Low-Temperature	Enhanced Photosynthesis	[44]
Oryza sativa	GB N-Acc.	Salinity, Chilling stress	Improve photosynthesis and phenotype	[45]
Gossypium hirsutum	GB Acc.	Drought	Osmotic adjustment, enhance yield	[46]
Nicotiana tabacum	GB N-Acc.	Salinity	Phenotypic traits	[47]
Triticum aestivum	GB Acc.	Heat and drought stress	Promoted photosynthesis, antioxidant and water status	[48]
Lycopersicum esculentum	GB N-Acc.	Salinity	Protect photosynthesis and reproductive organs	[49]
Lycopersicum esculentum	GB N-Acc.	High temperature	Enhanced the expression of heat-shock genes	[50]
Oryza sativa	GB N-Acc.	Water stress	Enhance Survival rate and agronomic traits	[51]
Lycopersicum esculentum	GB N-Acc.	Chilling stress	Promoted ROS scavenge	[52]
Brassica chinensis	GB N-Acc.	High salinity and high temperature	Promote photosynthesis	[53]

Table 1.
Reported roles of GB in transgenic plant under abiotic stresses.

produce a low concentration of GB that might not be sufficient for osmoregulation to counteract with abiotic stress [56].

The genes (codA or cDNA BADH) and enzymes involve in GB biosynthesis have been identified and cloned. GB has been successfully synthesized in various targeted organisms and provided stress tolerance via genetic engineering (**Table 2**).

2.3.1 Genetic engineering of GB via codA gene

As shown in **Table 2**, many species that can accumulate or not GB have been targeted via genetic engineering to synthesize or over accumulate GB under both stressed and non-stressed conditions. The choline oxidase (codA) from *A. globiformis* has been widely used in various transgenic plant species to synthesize GB, and *codA* has the ability to convert choline in one reaction [56].

The catalytic activity of choline oxidase (EC: 1.1.3.17) in *A. globiformis* results in this following equation: (Choline + H_2O + 2 O_2 = glycine betaine + H^+ + 2 H_2O_2) [63].

Transgenic species	GB Acc./ GB N-Acc.	Genes targeted	Protein Encoded	Organism sources/ Promoter	Roles in plant	References
Oryza sativa	GB N-Acc.	codA	Choline oxidase	*Arthrobacter globiformis*	Water stress tolerance	[51]
Zea mays/ Glycine max	GB Acc.	GB1(novel gene)	GB1 protein	*Zea mays* H-GB genotype / - *Agrobacterium* - Rice actin and - 35S promoter	Enhanced endogenous GB synthesis	[57]
Nicotiana tabacum	GB N-Acc.	cDNA sequence	BADH	*Spinacia oleracea* and *Beta vulgaris*	Betaine aldehyde resistance	[13]
Lycopersicum esculentum	GB N-Acc.	codA	Choline oxidase	*Arthrobacter globiformis*	Modulation of phosphate homeostasis under stress	[58]
Lycopersicum esculentum	GB N-Acc.	codA	Choline oxidase	*Arthrobacter globiformis*	Reproductive organs regulation	[59]
Brassica juncea	GB Acc.	codA	Choline oxidase	*Arthrobacter globiformis*	Photo inhibition tolerance	[11]
Nicotiana tabacum	GB N-Acc.	BADH cDNA	BADH	*Hordeum vulgare*	GB synthesis in non accumulator plant	[60]
Nicotiana tabacum	GB N-Acc.	BADH cDNA	BADH	*Escherichia coli*	Salt tolerance	[47]
Eucalyptus camaldulensis	GB Acc.	codA	Choline oxidase	*Arthrobacter globiformis*/ CaMV 35 promoter	Enhance of GB biosynthesis	[61]
Eucalyptus globulus	GB Acc.	codA	Choline oxidase	*Arthrobacter globiformis*	GB accumulation	[62]
Triticum aestivum	GB Acc.	BADH gene	BADH	*Atriplex hortensis*	Stress tolerance	[48]

Table 2.
Overview of GB genetic engineering in various plant species.

The codA gene is of particular interest with respect to the engineering of desirable productive traits in crop plants and stress tolerance. In transgenic tomato and brown mustard the codA was targeted to the chloroplast and cytosol which allowed GB accumulation for an increase of stress tolerance [19, 59]. Further, transgenic *indica* rice showed a significant increase of water-stress tolerance and transcriptome changes via codA gene expression [51]. One of the advantages of using choline oxidase pathway as a tool for engineering GB synthesis in plant is that the addition of a single gene codA is enough for the conversion of choline to GB [8]. The codA transgenic plant has showed their abilities to counteract with environmental stresses such as salinity, high temperature, high light, cold stress and freezing in different plant growth stages [64].

2.3.2 Genetic engineering of GB via BADH gene

The other pathway that provided successful results in genetic engineering of GB biosynthesis in various transgenic plant species is the BADH pathway (**Table 2**). BADH is one of the most prominent genes involved in the biosynthetic pathway of GB, and its utilization in various plant species has led to an increased tolerance to a variety of environmental stresses [65]. Indeed, the second step of GB biosynthesis is performed by betaine aldehyde dehydrogenase (BADH) that can be encoded by *betB* or *betA* gene from *E. coli*. BADH is an NDA-dependent dehydrogenase that has been characterized and cloned from plants species belong to the Amaranthaceae and Gramineae families [15]. The BADH pathway has been targeted in the chloroplasts in *N. tabacum* [13] and in peroxisomes in Gramineae [60]. Many studies showed positive results in stress tolerance in transgenic plants with genes *betA*, *betB* or both from *Escherichia coli* encoding Oxygen-dependent choline dehydrogenase (CHDH) and BADH [8]. The catalytic activities of CHDH (EC: 1.1.99.1) encode by *betA* from *E. coli* can be resume by this following Eq. (A + choline = AH2 + betaine aldehyde), A (hydrogen acceptor) and AH2 (hydrogen donor) [66]. Meanwhile the catalytic activities of the NAD/NADP-dependent betaine aldehyde dehydrogenase (EC: 1.2.1.8) are done by this equation: (betaine aldehyde + H2O + NAD^+ = glycine betaine +2 H^+ + NADH) [66, 67]. The equation for the catalytic activities is similar for chloroplastic betaine aldehyde dehydrogenase in sugar beet or spinach compared to those of *E. coli*.

3. Metabolism engineering of melatonin

Melatonin (**Figure 3**) as an ancient pleiotropic bio-molecule which can be traced back to the origin of life, is present in both animal and plant organisms [24, 68]. In plant, melatonin has been found in diverse family and at different stage of growth: Asteraceae, papaveracea, apiaceae, linaceae, fabaceae, poaceae, rosaceae, lamiaceae, solanaceae, musaceae or vitacea etc. [69].

Melatonin (N-acetyl-5-methoxytryptamine), a multifunctional plant hormone, was discovered in plants in 1995 [70]. Moreover, the presence of melatonin in plant was confirmed in *Chenopodium rubrum* via chromatography/tandem mass spectrometry and radio-immuno-assays [71]. Melatonin has multi-functional actions

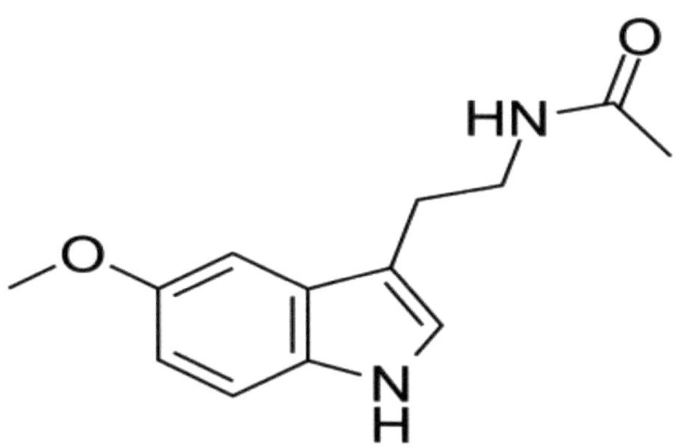

Figure 3.
N-acetyl-5-methoxytryptamine.

that improve cellular and organ health in various plant species and it is a powerful antioxidant in both animals and plants [72].

Melatonin functions as a metabolite with numerous roles in plant, including plant stress responses such as chilling, oxidative stress, drought, salt stress and nutrients deficiency, moreover melatonin can regulates plant growth and development, such as root organogenesis, flowering, and senescence [9, 73, 74]. Plenty of studies have focused on the function and regulation of melatonin in transgenic plants because of its crucial role in plant regulation.

3.1 Melatonin biosynthesis pathways in plant

The **Figure 4** shows a schematic representation of the biosynthesis of MT, in which the tryptophan is synthesized via shikimic acid pathway that is also responsible for the synthesis of vitamins and aromatic amino acids such as phenylalanine and tyrosine. In plants, tryptophan is converted to Tryptamine via a reaction catalyzed by tryptophan decarboxylase (TDC) [75], and the production of serotonin from Tryptamine is activated by tryptamine 5-hydroxylase [76]. The formation of melatonin is preceded by two reactions from serotonin; the first reaction catalyzed by ASMT transform serotonin to 5-methoxytryptamine, and the last step is catalyzed by N-acetyltransferase [77].

As far as we know, there are 6 genes which are involved in plant melatonin biosynthesis: TDC, TPH, T5H, SNAT, ASMT, and COMT [68], and the keys enzymes they encoded are the; L-tryptophan decarboxylase, tryptophan hydroxylase, serotonin-N-acetyltransferase, N-acetylserotonin methyltransferase and hydroxyindole-O-methyltransferase [24].

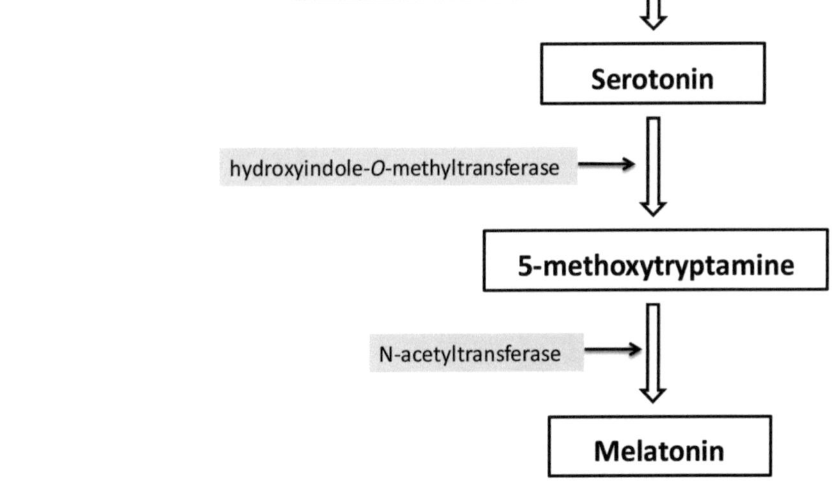

Figure 4.
A schematic representation of melatonin biosynthesis in brief.

3.2 Melatonin involve in abiotic stress tolerance

Melatonin is well know as a hormone which can significantly increase the plant survival rates, photosynthetic efficiency and antioxidant activities in plant under environmental stress [74, 78]. For these reasons, many studies were focused on the effects of exogenous melatonin on various plant species under abiotic stress. Indeed, exogenous melatonin could stimulate the biosynthesis of cold tolerance agents and contribute to increase the plant growth and development under cold stress [79]. As show **Table 3**, the alleviation of environmental stresses by melatonin has been investigated in many plant species: under drought (Zea mays) [89], under heavy metal (Caryaca thayensis) [90], under chilling stress (*Cynodon dactylon*) [91] and under salinity (*Cucumis sativus*) [82]. Compared to glycine betaine genetic engineering in plant under stress, the use of melatonin in transgenic plant to provide stress tolerance is fewer. However, there is several studies that focused on the over expression of melatonin via metabolic engineering (**Table 3**).

In Transgenic Arabidopsis the over expression of N-acetyltransferase gene increased salt tolerance via the increase in autophagy, and the reestablishment of redox and ion homeostasis [29]. Furthermore, increase of over-expressing N-acetyltransferase gene enhances the endogenous content in transgenic rice that provoked pleiotropic phenotypes, including enhanced seedling growth, delayed flowering, and low grain yield [28].

Plant species	Transgenic/exogenous	Stress	Role in stress	References
Oryza sativa	Transgenic	Chilling	Promote photosynthesis	[80]
Malus hupehensis	Exogenous MT	Salt stress	Boost antioxidant system	[81]
Arabidopsis thaliana	Transgenic	Drought	Enhanced melatonin content	[27]
Cucumis sativus	Exogenous MT	Salt stress	Enhanced the rate of germination	[82]
Lycopersicum esculentum	Transgenic	Drought	Enhanced melatonin content	[83]
Oryza sativa	Transgenic	Heavy metal stress (Cadmium)	Enhanced stress tolerance	[84]
Oryza sativa	Transgenic	Herbicide	oxidative stress resistance	[85]
Nicotiana sylvestris	Transgenic	UV-B radiation	Reduced DNA damages	[86]
Phacelia tanacetifolia	Exogenous MT	high temperature and light	Promoted germination	[87]
Lycopersicum esculentum	Transgenic	Salt stress	ROS scavenge	[88]
Arabidopsis thaliana	Transgenic	Salt stress	Increase in autophagy and rebalance homeostasis	[29]

Table 3.
Reported roles of MT exogenously applied and in transgenic plant under abiotic stresses.

Transgenic species	Genes targeted	Protein encoded	Organism source/ transformer/vector	Functions	References
Medicago sativa	*MsASMT*$_1$	N-acetylserotonin methyltransferase	*Alfalfa*/*Agrobacterium* strain EHA105/pZh01-MsASMT1 vector	Ameliorated Plant Growth	[92]
Panicum virgatum	*AANAT and HIOMT*	arylalkylamine N-acetyltransferase / hydroxyindole O-methyltransferase	Ovine/*Agrobacterium*-mediated method / vector Ubi1301	Improved growth and salt tolerance	[93]
Oryza sativa	*ASDAC*	N-acetylserotonin deacetylase	Rice/*Agrobacterium tumefaciens*/ pTCK303:ASDAC RNAi binary and pIPKb002:ASDAC vector	Regulation of melatonin in plant	[94]
Arabidopsis thaliana	*cDNA TaCOMT*	Caffeic acid 3-O-methyltransferase	Wheat/*Agrobacterium tumefaciens* strain GV3101 / pCAMBIA1302-TaCOMT vector	Promoted drought tolerance	[27]
Oryza sativa	*OaSNAT (SNAT)*	Serotonin N-acetyltransferase	Sheep/*Agrobacterium*-mediated method	Homeostasic regulation of melatonin	[95]
Arabidopsis thaliana	*MsSNAT*	serotonin N-acetyltransferase	Alfalfa/*Agrobacterium*-mediated method	Salt tolerance	[29]
Arabidopsis thaliana	*MzASMT*$_1$ *(ASTM)*	N-acetylserotonin-O-methyltransferase	Apple/35S promoter	Drought tolerance	[96]
Panicum virgatum	*HIOMT*	hydroxyindole O-methyltransferase	Ovine/*Agrobacterium*-mediated method	biosynthetic and physiological functional networks of melatonin	[97]
Nicotiana sylvestris	*AANAT HIOMT*	arylalkylamine N-acetyltransferase/ hydroxyindole-O-methyltransferase	*Agrobacterium tumefaciens*-mediated transformation	Inhibited UV-B-induced DNA damage	[86]
Lycopersicum esculentum	*SlCOMT1*	caffeic acid O-methyl-transferase	Tomato/*Agrobacterium* LBA4404/pMD18-T cloning, pCXSN-Myc, SlCOMT1-Myc over-expression vectors	Salt tolerance	[88]

Table 4.
Overview of MT metabolic engineering in diverse plants.

3.3 Melatonin in plant metabolism engineering

Previous studies using genetic engineering (transgenic plant) in various plants species with low or high MT accumulation has been achieved to determined the role of MT in plant growth regulation, stress tolerance or MT function in plant (**Table 4**). Indeed it was reported the implication of MT in seed germination, root development, fruit ripening, senescence, yield, circadian rhythm and plant homeostasis [98]. Ectopic over-expression (transgenesis) of human serotonin N-acetyltransferase increased endogenous melatonin that allowed transgenic rice seedlings to face chilling stress [80]. The increase of endogenous melatonin in various transgenic plant organisms compared to the wild type has been reported in *Arabidopsis thaliana* [29], in *Lycopersicum esculentum* [88] or in *Medicago sativa* [92].

Most of the studies in MT transgenesis are based on the ability of *Agrobacterium* to transfer DNA to plant cells by genetic engineering (**Table 4**). Indeed *Agrobacterium tumefaciens* is a widespread naturally occurring soil bacterium which demonstrated a great ability to introduce new genetic material into diverse plant cell species [99]. The *Agrobacterium*-mediated transformation process can be resumed in this following line: 1- Isolation of the targeted genes → 2- development of a functional transgenic construct → 3- insertion of the transgene → 4- introduction of the T-DNA-containing-plasmid into *Agrobacterium* → 5- mixture of the transformed *Agrobacterium* with plant cells → 6- regeneration of the transformed cells into transgenic plant → 7- testing for trait performance or transgene expression [99–101]. The catalytic activities of different enzymes involved in MT metabolic engineering have been elucidated in various species. The catalytic activity of Acetylserotonin O-methyltransferase (EC: 2.1.1.4) encoded by *ASMT* gene in *Homo sapiens* is done by this following line: (*N*-acetylserotonin + *S*-adenosyl-L-methionine = H^+ + melatonin + *S*-adenosyl-L-homocysteine) [102]. The catalytic activity of Serotonin N-acetyltransferase (EC: 2.3.1.87) from *Ovis aries* (Sheep) encoded by *AANAT* gene is done by this reaction: (2-arylethylamine + acetyl-CoA = CoA + H^+ + *N*-acetyl-2-arylethylamine) [103]. Moreover the catalytic activity of Caffeic acid 3-O-methyltransferase (EC: 2.1.1.68) implicated in many MT genetic engineering manipulations has been decoded in Medicago sativa (Alfalfa): ((*E*)-caffeate + *S*-adenosyl-L-methionine = (*E*)-ferulate + H^+ + *S*-adenosyl-L-homocysteine) [104].

The elucidations of these reactions and techniques provided a huge benefit to increase the use of those compounds in metabolic engineering. There are others areas to explore and clarify to shed light the use of melatonin or glycine betaine metabolic engineering.

Author details

Cisse El Hadji Malick*, Miao Ling-Feng, Li Da-Dong and Yang Fan
Key Laboratory of Agro-Forestry Environmental Processes and Ecological Regulation of Hainan Province, Center for Eco-Environmental Restoration Engineering of Hainan Province, School of Ecological and Environmental Sciences, Hainan University, Haikou, Hainan, China

*Address all correspondence to: kosmosofmalick@gmail.com

References

[1] Francini and Sebastiani, "Abiotic Stress Effects on Performance of Horticultural Crops," Horticulturae, vol. 5, no. 4, Sep. 2019, doi: 10.3390/horticulturae5040067.

[2] T. Khan, M. A. Khan, K. Karam, N. Ullah, Z.-R. Mashwani, and A. Nadhman, "Plant in vitro Culture Technologies; A Promise Into Factories of Secondary Metabolites Against COVID-19," *Front. Plant Sci.*, vol. 12, Mar. 2021, doi: 10.3389/fpls.2021.610194.

[3] Ganjewala D., Kaur G., Srivastava N. Metabolic Engineering of Stress Protectant Secondary Metabolites to Confer Abiotic Stress Tolerance in Plants. In: Singh S., Upadhyay S., Pandey A., Kumar S. (eds) Molecular Approaches in Plant Biology and Environmental Challenges. Energy, Environment, and Sustainability. 2019, Springer, Singapore. https://doi.org/10.1007/978-981-15-0690-1_11

[4] S. D. McNeil, M. L. Nuccio, and A. D. Hanson, "Betaines and Related Osmoprotectants. Targets for Metabolic Engineering of Stress Resistance," *Plant Physiol.*, vol. 120, no. 4, Aug. 1999, doi: 10.1104/pp.120.4.945.

[5] S. H. Wani, V. Kumar, V. Shriram, and S. K. Sah, "Phytohormones and their metabolic engineering for abiotic stress tolerance in crop plants," *Crop J.*, vol. 4, no. 3, Jun. 2016, doi: 10.1016/j.cj.2016.01.010.

[6] A. S. Birchfield and C. A. McIntosh, "Metabolic engineering and synthetic biology of plant natural products – A minireview," *Curr. Plant Biol.*, vol. 24, Dec. 2020, doi: 10.1016/j.cpb.2020.100163.

[7] S. S. Gill and N. Tuteja, "Polyamines and abiotic stress tolerance in plants," *Plant Signal. Behav.*, vol. 5, no. 1, Jan. 2010, doi: 10.4161/psb.5.1.10291.

[8] M. G. Annunziata, L. F. Ciarmiello, P. Woodrow, E. Dell'Aversana, and P. Carillo, "Spatial and Temporal Profile of Glycine Betaine Accumulation in Plants Under Abiotic Stresses," *Front. Plant Sci.*, vol. 10, Mar. 2019, doi: 10.3389/fpls.2019.00230.

[9] J. Li, J. Liu, T. Zhu, C. Zhao, L. Li, and M. Chen, "The Role of Melatonin in Salt Stress Responses," *Int. J. Mol. Sci.*, vol. 20, no. 7, Apr. 2019, doi: 10.3390/ijms20071735.

[10] G. Blunden, S. M. Gordon, T. A. Crabb, O. G. Roch, M. G. Rowan, and B. Wood, "NMR spectra of betaines from marine algae," *Magn. Reson. Chem.*, vol. 24, no. 11, Nov. 1986, doi: 10.1002/mrc.1260241108.

[11] K. V. S. . Prasad and P. P. Saradhi, "Enhanced tolerance to photoinhibition in transgenic plants through targeting of glycinebetaine biosynthesis into the chloroplasts," *Plant Sci.*, vol. 166, no. 5, May 2004, doi: 10.1016/j.plantsci.2003.12.031.

[12] D. Rhodes and A. D. Hanson, "Quaternary Ammonium and Tertiary Sulfonium Compounds in Higher Plants," *Annu. Rev. Plant Physiol. Plant Mol. Biol.*, vol. 44, no. 1, Jun. 1993, doi: 10.1146/annurev.pp.44.060193.002041.

[13] B. Rathinasabapathi, K. McCue, D. Gage, and A. Hanson, "Metabolic engineering of glycine betaine synthesis: plant betaine aldehyde dehydrogenases lacking typical transit peptides are targeted to tobacco chloroplasts where they confer betaine aldehyde resistance," *Planta*, vol. 193, no. 2, Mar. 1994, doi: 10.1007/BF00192524.

[14] E. A. Weretilnyk and A. D. Hanson, "Molecular cloning of a plant betaine-aldehyde dehydrogenase, an enzyme implicated in adaptation to salinity and drought.," *Proc. Natl. Acad. Sci.*, vol. 87,

no. 7, Apr. 1990, doi: 10.1073/pnas.87.7.2745.

[15] S. D. McNeil, M. L. Nuccio, and A. D. Hanson, "Betaines and related osmoprotectants. Targets for metabolic engineering of stress resistance," *Plant Physiol.*, vol. 120, no. 4, pp. 945-949, 1999, doi: 10.1104/pp.120.4.945.

[16] D. Le Rudulier, A. R. Strom, A. M. Dandekar, L. T. Smith, and R. C. Valentine, "Molecular Biology of Osmoregulation," *Science (80-.).*, vol. 224, no. 4653, Jun. 1984, doi: 10.1126/science.224.4653.1064.

[17] Z. Xu *et al.*, "Glycinebetaine Biosynthesis in Response to Osmotic Stress Depends on Jasmonate Signaling in Watermelon Suspension Cells," *Front. Plant Sci.*, vol. 9, Oct. 2018, doi: 10.3389/fpls.2018.01469.

[18] A. Sakamoto and N. Murata, "Genetic engineering of glycinebetaine synthesis in plants: current status and implications for enhancement of stress tolerance," *J. Exp. Bot.*, vol. 51, no. 342, Jan. 2000, doi: 10.1093/jexbot/51.342.81.

[19] H. Hayashi, Alia, L. Mustardy, P. Deshnium, M. Ida, and N. Murata, "Transformation of Arabidopsis thaliana with the codA gene for choline oxidase; accumulation of glycinebetaine and enhanced tolerance to salt and cold stress," *Plant J.*, vol. 12, no. 1, Jul. 1997, doi: 10.1046/j.1365-313X.1997.12010133.x.

[20] M. S. Khan, X. Yu, A. Kikuchi, M. Asahina, and K. N. Watanabe, "Genetic engineering of glycine betaine biosynthesis to enhance abiotic stress tolerance in plants," *Plant Biotechnol.*, vol. 26, no. 1, 2009, doi: 10.5511/plantbiotechnology.26.125.

[21] Y. Yu *et al.*, "The Role of Phyto-Melatonin and Related Metabolites in Response to Stress," *Molecules*, vol. 23, no. 8, Jul. 2018, doi: 10.3390/molecules23081887.

[22] M. B. Arnao and J. Hernández-Ruiz, "Melatonin and its relationship to plant hormones," *Ann. Bot.*, vol. 121, no. 2, Feb. 2018, doi: 10.1093/aob/mcx114.

[23] W. Ma, L. Xu, S. Gao, X. Lyu, X. Cao, and Y. Yao, "Melatonin alters the secondary metabolite profile of grape berry skin by promoting VvMYB14-mediated ethylene biosynthesis," *Hortic. Res.*, vol. 8, no. 1, Dec. 2021, doi: 10.1038/s41438-021-00478-2.

[24] D. Zhao *et al.*, "Melatonin Synthesis and Function: Evolutionary History in Animals and Plants," *Front. Endocrinol. (Lausanne).*, vol. 10, Apr. 2019, doi: 10.3389/fendo.2019.00249.

[25] J. Axelrod and H. Weissbach, "Enzymatic O-Methylation of N-Acetylserotonin to Melatonin," *Science (80-.).*, vol. 131, no. 3409, Apr. 1960, doi: 10.1126/science.131.3409.1312.

[26] Y. Byeon, H. Y. Lee, K. Lee, and K. Back, "Caffeic acid *O* -methyltransferase is involved in the synthesis of melatonin by methylating *N* -acetylserotonin in *Arabidopsis*," *J. Pineal Res.*, vol. 57, no. 2, Sep. 2014, doi: 10.1111/jpi.12160.

[27] W.-J. Yang *et al.*, "Overexpression of TaCOMT Improves Melatonin Production and Enhances Drought Tolerance in Transgenic Arabidopsis," *Int. J. Mol. Sci.*, vol. 20, no. 3, Feb. 2019, doi: 10.3390/ijms20030652.

[28] Y. Byeon and K. Back, "An increase in melatonin in transgenic rice causes pleiotropic phenotypes, including enhanced seedling growth, delayed flowering, and low grain yield," *J. Pineal Res.*, vol. 56, no. 4, May 2014, doi: 10.1111/jpi.12129.

[29] G. Zhao *et al.*, "Transgenic Arabidopsis overexpressing MsSNAT enhances salt tolerance via the increase

in autophagy, and the reestablishment of redox and ion homeostasis," *Environ. Exp. Bot.*, vol. 164, Aug. 2019, doi: 10.1016/j.envexpbot.2019.04.017.

[30] M. L. Nuccio, S. D. McNeil, M. J. Ziemak, A. D. Hanson, R. K. Jain, and G. Selvaraj, "Choline Import into Chloroplasts Limits Glycine Betaine Synthesis in Tobacco: Analysis of Plants Engineered with a Chloroplastic or a Cytosolic Pathway," *Metab. Eng.*, vol. 2, no. 4, Oct. 2000, doi: 10.1006/mben.2000.0158.

[31] S. D. McNeil, M. L. Nuccio, M. J. Ziemak, and A. D. Hanson, "Enhanced synthesis of choline and glycine betaine in transgenic tobacco plants that overexpress phosphoethanolamine N-methyltransferase," *Proc. Natl. Acad. Sci.*, vol. 98, no. 17, Aug. 2001, doi: 10.1073/pnas.171228998.

[32] D. Rontein *et al.*, "Plants Synthesize Ethanolamine by Direct Decarboxy lation of Serine Using a Pyridoxal Phosphate Enzyme," *J. Biol. Chem.*, vol. 276, no. 38, Sep. 2001, doi: 10.1074/jbc.M106038200.

[33] P. S. Summers and E. A. Weretilnyk, "Choline Synthesis in Spinach in Relation to Salt Stress," *Plant Physiol.*, vol. 103, no. 4, Dec. 1993, doi: 10.1104/pp.103.4.1269.

[34] T. L. Fitzgerald, D. L. E. Waters, and R. J. Henry, "Betaine aldehyde dehydrogenase in plants," *Plant Biol.*, vol. 11, no. 2, Mar. 2009, doi: 10.1111/j.1438-8677.2008.00161.x.

[35] T. Demiral and I. Türkan, "Does exogenous glycinebetaine affect antioxidative system of rice seedlings under NaCl treatment?," *J. Plant Physiol.*, vol. 161, no. 10, Oct. 2004, doi: 10.1016/j.jplph.2004.03.009.

[36] P. Malekzadeh, "Influence of exogenous application of glycinebetaine on antioxidative system and growth of salt-stressed soybean seedlings (Glycine max L.)," *Physiol. Mol. Biol. Plants*, vol. 21, no. 2, Apr. 2015, doi: 10.1007/s12298-015-0292-4.

[37] Y. Wang and N. Nii, "Changes in chlorophyll, ribulose bisphosphate carboxylase-oxygenase, glycine betaine content, photosynthesis and transpiration in *Amaranthus tricolor* leaves during salt stress," *J. Hortic. Sci. Biotechnol.*, vol. 75, no. 6, Jan. 2000, doi: 10.1080/14620316.2000.11511297.

[38] S. Kishitani, K. Watanabe, S. Yasuda, K. Arakawa, and T. Takabe, "Accumulation of glycinebetaine during cold acclimation and freezing tolerance in leaves of winter and spring barley plants," *Plant, Cell Environ.*, vol. 17, no. 1, Jan. 1994, doi: 10.1111/j.1365-3040.1994.tb00269.x.

[39] Q.-Q. Ma, W. Wang, Y.-H. Li, D.-Q. Li, and Q. Zou, "Alleviation of photoinhibition in drought-stressed wheat (Triticum aestivum) by foliar-applied glycinebetaine," *J. Plant Physiol.*, vol. 163, no. 2, Feb. 2006, doi: 10.1016/j.jplph.2005.04.023.

[40] S. H. Raza, H. R. Athar, M. Ashraf, and A. Hameed, "Glycinebetaine-induced modulation of antioxidant enzymes activities and ion accumulation in two wheat cultivars differing in salt tolerance," *Environ. Exp. Bot.*, vol. 60, no. 3, Jul. 2007, doi: 10.1016/j.envexpbot.2006.12.009.

[41] M. A. . Raza, M. . Saleem, G. . Shah, I. . Khan, and A. Raza, "Exogenous application of glycinebetaine and potassium for improving water relations and grain yield of wheat under drought," *J. soil Sci. plant Nutr.*, no. ahead, 2014, doi: 10.4067/S0718-95162014005000028.

[42] K. O. Holmström, S. Somersalo, A. Mandal, T. E. Palva, and B. Welin, "Improved tolerance to salinity and low temperature in transgenic tobacco

producing glycine betaine," *J. Exp. Bot.*, vol. 51, no. 343, pp. 177-185, 2000, doi: 10.1093/jexbot/51.343.177.

[43] R. Quan, M. Shang, H. Zhang, Y. Zhao, and J. Zhang, "Improved chilling tolerance by transformation with betA gene for the enhancement of glycinebetaine synthesis in maize," *Plant Sci.*, vol. 166, no. 1, Jan. 2004, doi: 10.1016/j.plantsci.2003.08.018.

[44] P. Deshnium, Z. Gombos, Y. Nishiyama, and N. Murata, "The action in vivo of glycine betaine in enhancement of tolerance of Synechococcus sp. strain PCC 7942 to low temperature.," *J. Bacteriol.*, vol. 179, no. 2, 1997, doi: 10.1128/JB.179.2.339-344.1997.

[45] A. Sakamoto and A. and N. Murata, "Metabolic engineering of rice leading to biosynthesis of glycinebetaine and tolerance to salt and cold," *Plant Mol. Biol.*, vol. 38, no. 6, 1998, doi: 10.1023/A:1006095015717.

[46] S. Lv, A. Yang, K. Zhang, L. Wang, and J. Zhang, "Increase of glycinebetaine synthesis improves drought tolerance in cotton," *Mol. Breed.*, vol. 20, no. 3, Aug. 2007, doi: 10.1007/s11032-007-9086-x.

[47] G. Lilius, N. Holmberg, and L. Bülow, "Enhanced NaCl Stress Tolerance in Transgenic Tobacco Expressing Bacterial Choline Dehydrogenase," *Nat. Biotechnol.*, vol. 14, no. 2, Feb. 1996, doi: 10.1038/nbt0296-177.

[48] G. P. Wang, X. Y. Zhang, F. Li, Y. Luo, and W. Wang, "Overaccumulation of glycine betaine enhances tolerance to drought and heat stress in wheat leaves in the protection of photosynthesis," *Photosynthetica*, vol. 48, no. 1, Mar. 2010, doi: 10.1007/s11099-010-0016-5.

[49] E.-J. Park, Z. Jeknic, M.-T. Pino, N. Murata, and T. H.-H. Chen, "Glycinebetaine accumulation is more effective in chloroplasts than in the cytosol for protecting transgenic tomato plants against abiotic stress," *Plant. Cell Environ.*, vol. 30, no. 8, Aug. 2007, doi: 10.1111/j.1365-3040.2007.01694.x.

[50] S. LI *et al.*, "Glycinebetaine enhances the tolerance of tomato plants to high temperature during germination of seeds and growth of seedlings," *Plant. Cell Environ.*, vol. 34, no. 11, Nov. 2011, doi: 10.1111/j.1365-3040.2011.02389.x.

[51] H. Kathuria, J. Giri, K. N. Nataraja, N. Murata, M. Udayakumar, and A. K. Tyagi, "Glycinebetaine-induced water-stress tolerance in *codA* -expressing transgenic *indica* rice is associated with up-regulation of several stress responsive genes," *Plant Biotechnol. J.*, vol. 7, no. 6, Aug. 2009, doi: 10.1111/j.1467-7652.2009.00420.x.

[52] E.-J. Park *et al.*, "Genetic engineering of glycinebetaine synthesis in tomato protects seeds, plants, and flowers from chilling damage," *Plant J.*, vol. 40, no. 4, Oct. 2004, doi: 10.1111/j.1365-313X.2004.02237.x.

[53] Q. Wang, W. Xu, Q. Xue, and W. Su, "Transgenic Brassica chinensis plants expressing a bacterial codA gene exhibit enhanced tolerance to extreme temperature and high salinity," *J. Zhejiang Univ. Sci. B*, vol. 11, no. 11, Nov. 2010, doi: 10.1631/jzus.B1000137.

[54] P. Yancey, "Compatible and counteracting solutes, in cellular and molecular physiology," *Acad. Press*, 1994.

[55] H. J. M. Hou *et al.*, "Inducers of glycinebetaine synthesis in barley," *Plant Physiol.*, vol. 120, no. 4, pp. 856-865, 2019, doi: 10.1104/pp.120.4.945.

[56] J. Giri, "Glycinebetaine and abiotic stress tolerance in plants," *Plant Signal. Behav.*, vol. 6, no. 11, pp. 1746-1751, 2011, doi: 10.4161/psb.6.11.17801.

[57] P. Castiglioni *et al.*, "Identification of *GB1*, a gene whose constitutive overexpression increases glycinebetaine content in maize and soybean," *Plant Direct*, vol. 2, no. 2, Feb. 2018, doi: 10.1002/pld3.40.

[58] D. Li *et al.*, "Genetic Engineering of the Biosynthesis of Glycine Betaine Modulates Phosphate Homeostasis by Regulating Phosphate Acquisition in Tomato," *Front. Plant Sci.*, vol. 9, Jan. 2019, doi: 10.3389/fpls.2018.01995.

[59] E.-J. Park, Z. Jeknić, T. H. H. Chen, and N. Murata, "The codA transgene for glycinebetaine synthesis increases the size of flowers and fruits in tomato," *Plant Biotechnol. J.*, vol. 5, no. 3, May 2007, doi: 10.1111/j.1467-7652.2007.00251.x.

[60] T. Nakamura *et al.*, "Expression of a betaine aldehyde dehydrogenase gene in rice, a glycinebetaine nonaccumulator, and possible localization of its protein in peroxisomes," *Plant J.*, vol. 11, no. 5, May 1997, doi: 10.1046/j.1365-313X.1997.11051115.x.

[61] N.-H. T. Tran, T. Oguchi, E. Matsunaga, A. Kawaoka, K. N. Watanabe, and A. Kikuchi, "Transcriptional enhancement of a bacterial *choline oxidase A* gene by an *HSP* terminator improves the glycine betaine production and salinity stress tolerance of *Eucalyptus camaldulensis* trees," *Plant Biotechnol.*, vol. 35, no. 3, Sep. 2018, doi: Arthrobacter globiformis.

[62] E. Matsunaga *et al.*, "Agrobacterium-mediated transformation of Eucalyptus globulus using explants with shoot apex with introduction of bacterial choline oxidase gene to enhance salt tolerance," *Plant Cell Rep.*, vol. 31, no. 1, Jan. 2012, doi: 10.1007/s00299-011-1159-y.

[63] T. Rand, T. Halkier, and O. C. Hansen, "Structural Characterization and Mapping of the Covalently Linked FAD Cofactor in Choline Oxidase from *Arthrobacter globiformis*," *Biochemistry*, vol. 42, no. 23, Jun. 2003, doi: 10.1021/bi0274266.

[64] S. Hussain Wani, N. Brajendra Singh, A. Haribhushan, and J. Iqbal Mir, "Compatible Solute Engineering in Plants for Abiotic Stress Tolerance - Role of Glycine Betaine," *Curr. Genomics*, vol. 14, no. 3, Apr. 2013, doi: 10.2174/1389202911314030001.

[65] M. Niazian, S. A. Sadat-Noori, M. Tohidfar, S. M. M. Mortazavian, and P. Sabbatini, "Betaine Aldehyde Dehydrogenase (BADH) vs. Flavodoxin (Fld): Two Important Genes for Enhancing Plants Stress Tolerance and Productivity," *Front. Plant Sci.*, vol. 12, Apr. 2021, doi: 10.3389/fpls.2021.650215.

[66] B. Landfald and A. R. Strøm, "Choline-glycine betaine pathway confers a high level of osmotic tolerance in Escherichia coli.," *J. Bacteriol.*, vol. 165, no. 3, 1986, doi: 10.1128/JB.165.3.849-855.1986.

[67] P. Falkenberg and A. R. Strøm, "Purification and characterization of osmoregulatory betaine aldehyde dehydrogenase of Escherichiacoli," *Biochim. Biophys. Acta - Gen. Subj.*, vol. 1034, no. 3, Jun. 1990, doi: 10.1016/0304-4165(90)90046-Y.

[68] K. Back, "Melatonin metabolism, signaling and possible roles in plants," *Plant J.*, vol. 105, no. 2, Jan. 2021, doi: 10.1111/tpj.14915.

[69] M. A. Nawaz *et al.*, "Melatonin: Current Status and Future Perspectives in Plant Science," *Front. Plant Sci.*, vol. 6, Jan. 2016, doi: 10.3389/fpls.2015.01230.

[70] R. Dubbels *et al.*, "Melatonin in edible plants identified by radioimmunoassay and by high performance liquid chromatography-mass spectrometry," *J. Pineal Res.*, vol.

18, no. 1, Jan. 1995, doi: 10.1111/j.1600-079X.1995.tb00136.x.

[71] J. Kolář *et al.*, "Melatonin: Occurrence and daily rhythm in *Chenopodium rubrum*," *Phytochemistry*, vol. 44, no. 8, Apr. 1997, doi: 10.1016/S0031-9422(96)00568-7.

[72] G. Cheng *et al.*, "Plant-derived melatonin from food: a gift of nature," *Food Funct.*, vol. 12, no. 7, 2021, doi: 10.1039/D0FO03213A.

[73] R. Sharif *et al.*, "Melatonin and its effects on plant systems," *Molecules*, vol. 23, no. 9, pp. 1-20, 2018, doi: 10.3390/molecules23092352.

[74] M. B. Arnao and J. Hernández-Ruiz, "Melatonin: A New Plant Hormone and/or a Plant Master Regulator?," *Trends Plant Sci.*, vol. 24, no. 1, Jan. 2019, doi: 10.1016/j.tplants.2018.10.010.

[75] V. De Luca, C. Marineau, and N. Brisson, "Molecular cloning and analysis of cDNA encoding a plant tryptophan decarboxylase: comparison with animal dopa decarboxylases.," *Proc. Natl. Acad. Sci.*, vol. 86, no. 8, Apr. 1989, doi: 10.1073/pnas.86.8.2582.

[76] M. PARK, K. KANG, S. PARK, and K. BACK, "Conversion of 5-Hydroxytryptophan into Serotonin by Tryptophan Decarboxylase in Plants, *Escherichia coli* , and Yeast," *Biosci. Biotechnol. Biochem.*, vol. 72, no. 9, Sep. 2008, doi: 10.1271/bbb.80220.

[77] D. Zhao, R. Wang, D. Liu, Y. Wu, J. Sun, and J. Tao, "Melatonin and Expression of Tryptophan Decarboxylase Gene (TDC) in Herbaceous Peony (*Paeonia lactiflora* Pall.) Flowers," *Molecules*, vol. 23, no. 5, May 2018, doi: 10.3390/molecules23051164.

[78] E. H. M. Cisse, L.-F. Miao, F. Yang, J.-F. Huang, D.-D. Li, and J. Zhang, "Gly Betaine Surpasses Melatonin to Improve Salt Tolerance in Dalbergia odorifera," *Front. Plant Sci.*, vol. 12, Feb. 2021.

[79] H. Shi *et al.*, "Comparative physiological, metabolomic, and transcriptomic analyses reveal mechanisms of improved abiotic stress resistance in bermudagrass [*Cynodon dactylon* (L). Pers.] by exogenous melatonin," *J. Exp. Bot.*, vol. 66, no. 3, Feb. 2015, doi: 10.1093/jxb/eru373.

[80] K. Kang, K. Lee, S. Park, Y. S. Kim, and K. Back, "Enhanced production of melatonin by ectopic overexpression of human serotonin N-acetyltransferase plays a role in cold resistance in transgenic rice seedlings," *J. Pineal Res.*, Jun. 2010, doi: 10.1111/j.1600-079X.2010.00783.x.

[81] C. Li *et al.*, "The mitigation effects of exogenous melatonin on salinity-induced stress in Malus hupehensis," *J. Pineal Res.*, vol. 53, no. 3, Oct. 2012, doi: 10.1111/j.1600-079X.2012.00999.x.

[82] H.-J. Zhang *et al.*, "Melatonin promotes seed germination under high salinity by regulating antioxidant systems, ABA and GA $_4$ interaction in cucumber (*Cucumis sativus* L.)," *J. Pineal Res.*, vol. 57, no. 3, Oct. 2014, doi: 10.1111/jpi.12167.

[83] L. Wang *et al.*, "Changes in melatonin levels in transgenic 'Micro-Tom' tomato overexpressing ovine *AANAT* and ovine *HIOMT* genes," *J. Pineal Res.*, vol. 56, no. 2, Mar. 2014, doi: 10.1111/jpi.12105.

[84] Y. Byeon, H. Y. Lee, O. J. Hwang, H.-J. Lee, K. Lee, and K. Back, "Coordinated regulation of melatonin synthesis and degradation genes in rice leaves in response to cadmium treatment," *J. Pineal Res.*, vol. 58, no. 4, May 2015, doi: 10.1111/jpi.12232.

[85] S. Park, D.-E. Lee, H. Jang, Y. Byeon, Y.-S. Kim, and K. Back, "Melatonin-rich transgenic rice plants exhibit resistance

to herbicide-induced oxidative stress," *J. Pineal Res.*, vol. 54, no. 3, Apr. 2013, doi: 10.1111/j.1600-079X.2012.01029.x.

[86] L. Zhang, *J. jia*, Y. Xu, Y. Wang, J. Hao, and T. Li, "Production of transgenic Nicotiana sylvestris plants expressing melatonin synthetase genes and their effect on UV-B-induced DNA damage," *Vitr. Cell. Dev. Biol. - Plant*, vol. 48, no. 3, Jun. 2012, doi: 10.1007/s11627-011-9413-0.

[87] I. Tiryaki and H. Keles, "Reversal of the inhibitory effect of light and high temperature on germination of Phacelia tanacetifolia seeds by melatonin," *J. Pineal Res.*, vol. 52, no. 3, Apr. 2012, doi: 10.1111/j.1600-079X.2011.00947.x.

[88] Liu, Sun, Liu, Shi, Chen, and Zhao, "Overexpression of the Melatonin Synthesis-Related Gene SlCOMT1 Improves the Resistance of Tomato to Salt Stress," *Molecules*, vol. 24, no. 8, Apr. 2019, doi: 10.3390/molecules24081514.

[89] X. Ma, J. Zhang, P. Burgess, *S. Rossi*, and B. Huang, "Interactive effects of melatonin and cytokinin on alleviating drought-induced leaf senescence in creeping bentgrass (*Agrostis stolonifera*)," *Environ. Exp. Bot.*, vol. 145, Jan. 2018, doi: 10.1016/j.envexpbot.2017.10.010.

[90] A. Sharma *et al.*, "Melatonin regulates the functional components of photosynthesis, antioxidant system, gene expression, and metabolic pathways to induce drought resistance in grafted Carya cathayensis plants," *Sci. Total Environ.*, vol. 713, Apr. 2020, doi: 10.1016/j.scitotenv.2020.136675.

[91] J. Fan *et al.*, "Alleviation of cold damage to photosystem II and metabolisms by melatonin in Bermudagrass," *Front. Plant Sci.*, vol. 6, Nov. 2015, doi: 10.3389/fpls.2015.00925.

[92] H. Cen *et al.*, "Overexpression of MsASMT1 Promotes Plant Growth and Decreases Flavonoids Biosynthesis in Transgenic Alfalfa (Medicago sativa L.)," *Front. Plant Sci.*, vol. 11, Apr. 2020, doi: 10.3389/fpls.2020.00489.

[93] Y.-H. Huang *et al.*, "Overexpression of ovine AANAT and HIOMT genes in switchgrass leads to improved growth performance and salt-tolerance," *Sci. Rep.*, vol. 7, no. 1, Dec. 2017, doi: 10.1038/s41598-017-12566-2.

[94] K. Lee, O. J. Hwang, and K. Back, "Rice N-acetylserotonin deacetylase regulates melatonin levels in transgenic rice," *Melatonin Res.*, vol. 3, no. 1, Mar. 2020, doi: 10.32794/mr11250046.

[95] Y. Byeon, H. Y. Lee, and K. Back, "Chloroplastic and cytoplasmic overexpression of sheep serotonin *N* -acetyltransferase in transgenic rice plants is associated with low melatonin production despite high enzyme activity," *J. Pineal Res.*, vol. 58, no. 4, May 2015, doi: 10.1111/jpi.12231.

[96] B. Zuo *et al.*, "Overexpression of *MzASMT* improves melatonin production and enhances drought tolerance in transgenic *Arabidopsis thaliana* plants," *J. Pineal Res.*, vol. 57, no. 4, Nov. 2014, doi: 10.1111/jpi.12180.

[97] S. Yuan *et al.*, "Comparative Transcriptomic Analyses of Differentially Expressed Genes in Transgenic Melatonin Biosynthesis Ovine HIOMT Gene in Switchgrass," *Front. Plant Sci.*, vol. 7, Nov. 2016, doi: 10.3389/fpls.2016.01613.

[98] H. Shi, K. Chen, Y. Wei, and C. He, "Fundamental Issues of Melatonin-Mediated Stress Signaling in Plants," *Front. Plant Sci.*, vol. 7, Jul. 2016, doi: 10.3389/fpls.2016.01124.

[99] S. B. Gelvin, "Agrobacterium-Mediated Plant Transformation: the Biology behind the 'Gene-Jockeying'

Tool," *Microbiol. Mol. Biol. Rev.*, vol. 67, no. 1, Mar. 2003, doi: 10.1128/MMBR.67.1.16-37.2003.

[100] K. K. Kumar, S. Maruthasalam, M. Loganathan, D. Sudhakar, and P. Balasubramanian, "An improved Agrobacterium-mediated transformation protocol for recalcitrant elite indica rice cultivars," *Plant Mol. Biol. Report.*, vol. 23, no. 1, Mar. 2005, doi: 10.1007/BF02772648.

[101] H. D. Jones, A. Doherty, and H. Wu, "Review of methodologies and a protocol for the Agrobacterium-mediated transformation of wheat," *Plant Methods*, vol. 1, no. 1, 2005, doi: 10.1186/1746-4811-1-5.

[102] H. G. Botros *et al.*, "Crystal structure and functional mapping of human ASMT, the last enzyme of the melatonin synthesis pathway," *J. Pineal Res.*, vol. 54, no. 1, Jan. 2013, doi: 10.1111/j.1600-079X.2012.01020.x.

[103] J. Pavlicek *et al.*, "Evidence That Proline Focuses Movement of the Floppy Loop of Arylalkylamine N-Acetyltransferase (EC 2.3.1.87)," *J. Biol. Chem.*, vol. 283, no. 21, May 2008, doi: 10.1074/jbc.M800593200.

[104] R. Edwards and R. A. Dixon, "Purification and characterization of S-adenosyl-l-methionine: Caffeic acid 3-O-methyltransferase from suspension cultures of alfalfa (Medicago sativa L.)," *Arch. Biochem. Biophys.*, vol. 287, no. 2, Jun. 1991, doi: 10.1016/0003-9861(91)90492-2.

Chapter 2

Soil Carbon Storage Potential of Tropical Grasses: A Review

Bezaye Gorfu Tessema, Heiko Daniel, Zenebe Adimassu and Brian Wilson

Abstract

Environmental degradation and climate change are key current threats to world agriculture and food security and human–induced changes have been significant driving forces of this global environmental change. An important component is land degradation which results in a diminished soil organic carbon (SOC) stock with concomitant loss of soil condition and function. Land management to improve soil organic matter content, condition and productivity is therefore a key strategy to safeguard agricultural production, food supply and environmental quality. Soil organic carbon sequestration through the use of plant species with high photosynthetic efficiency, deep roots and high biomass production is one important strategy to achieve this. Tropical pastures, which are adapted to a wide range of environmental conditions have particular potential in this regard and have been used extensively for land rehabilitation. Tropical pastures also have advantages over trees for biomass and carbon accumulation due to their rapid establishment, suitability for annual harvest, continual and rapid growth rates. In addition, tropical pastures have the potential for SOC storage in subsoil horizons due to their deep root systems and can be used as biomass energy crops, which could further promote their use as a climate change mitigation option. Here we aimed to review current knowledge regarding the SOC storage potential of tropical grasses worldwide and identified knowledge gaps and current research needs for the use of tropical grasses in agricultural production system.

Keywords: Soil organic carbon, Tropical perennial grass, climate change mitigation

1. Introduction

Environmental degradation and climate change are key current threats to world agriculture and food security [1–5]. Human–induced changes to land cover have been significant driving forces of this global environmental change, of which, soil degradation resulting from land conversion, agricultural intensification, soil disturbance and increased erosion have been key factors [6–9]. An important component of this land degradation globally has been a diminished SOC stock with concomitant loss of soil condition and function, compromising food production and agricultural sustainability [10–12]. Land and soil management to increase soil organic matter content, soil condition and productivity is therefore a key need globally to safeguard agricultural production, food supply and environmental quality.

Organic carbon in soils globally is estimated to be between 1500 and 1600 Gt [13, 14] to 1.0 m depth which represents a significant component of the global carbon cycle, storing more carbon than is contained in vegetation and the atmosphere combined [15–17]. It has been estimated that, worldwide, soils have lost between 42 and 78 Gt of their original SOC as a result of management pressures [18]. With this carbon depletion, however, comes a significant opportunity, since soils are believed to have the capacity to store an additional 0.4–1.2 Gt C year^{-1} with the introduction of more judicious land management practices [3, 7, 19–22]. As such, soils globally have considerable potential to offset greenhouse gas (GHG) emissions and SOC storage has been widely promoted as an important strategy to help meet national and international emissions reduction targets [23]. Additional SOC storage might therefore have the dual benefit of contributing to our response to climate change globally whilst helping to restore soil condition and function to promote sustainable land management, improved production and productivity [3, 7, 19, 20, 24].

Methodologies and management practices that reduce SOC loss or promote the storage of additional soil carbon are being actively investigated globally. It has been widely reported that cultivation accelerates organic matter decomposition by exposing sites within soil aggregates that were previously protected [25–29] while soil erosion, vegetation clearing and removal of crop residue are also known to result in long–term soil carbon loss [30, 31]. However, there are management practices which seem to either arrest SOC loss (e.*G. minimum* tillage) or to promote carbon storage such as afforestation, pasture conversion, grazing management, cover crops, water harvesting, erosion control and the use of soil amendments including biochar [32]. Not all of these are practical in production landscapes globally and not all will be equally effective in the management of SOC. The effectiveness of various management practices is therefore being explored to facilitate optimum carbon storage that can be integrated with agricultural production systems.

An approach that has attracted particular attention is the use of perennial grass species within the production system, which appear to significantly increase SOC across a range of environments and this is particularly true where these perennial grasses replace cropping systems [33–36]. Pastures are varied in terms of their geographical distribution and species composition comprising native and exotic, annual and perennial grasses, legumes, herbs and shrubs [37]. They are the primary resource for many farm industries and are the basis for the production of meat, wool, milk and fodder. Schuman et al. [38] estimated the SOC under grazing lands of the world to be 10–30% of the total global SOC stock, while Janssens et al. [39] estimated the overall C sink in grassland soils of most European countries to average approximately 60 g C m^{-2} year^{-1}.

Tropical perennial grass species have been particularly promoted due to the high biomass and carbon accumulation resulting from their excellent photosynthetic efficiency, rapid establishment, fast growth, deep root systems and potential annual harvest [28, 40, 41] and Parton et al. [42] suggested that tropical grasses have significant potential as a carbon sink. However, there is a research need to fully quantify their capacity to store additional soil carbon relative to other management systems and hence, their potential for greenhouse gas (GHG) abatement and soil condition recovery [43–45].

Here we aimed to review current knowledge with regard to the SOC storage potential of tropical grasses worldwide given their wide distribution and extensive use, where current agricultural policy environments have identified land management innovations as key entry point to achieve co-benefits of resilient agriculture, poverty alleviation, and climate change mitigation. Hence, we identify knowledge gaps and current research needs to fully explore the potential of tropical grass species for SOC change.

2. SOC storage potential of tropical grasses

A number of studies have considered the soil carbon storage potential of tropical pastures by comparison with other management systems. An empirical, five year study of tropical ecosystems in South America by Amézquita et al. [46], demonstrated that although tropical pastures were second only to native forest in the quantity of SOC stored, organic carbon in the soils of these pasture systems represented a higher proportion (95–98%) of the total ecosystem carbon than comparable native tropical forest systems and silvo–pastoral systems. Desjardins et al. [47], reported that where tropical forest was converted to tropical pasture in Brazilian Amazonia, a slight increase in SOC content occurred in both sandy and clay soils while Post and Kwon [48] described the similarity of the average rates of SOC accumulation in forest and grasslands of 33.8 and 33.2 g C m^{-2} y^{-1}, respectively through time following management although above ground carbon is lost. In Australia, Chan and McCoy [43] also identified the potential of introduced perennial pasture (Kikuyu) to store a mean of 73 Mg C ha^{-1} in soil which was similar to soils under native trees (77 Mg ha^{-1}). Under some circumstances, tropical pastures have been reported to have a greater capacity to store SOC compared with trees or forest. For example,

Grass type	Age (Year)	Sampling depth (cm)	Mg C ha^{-1} yr^{-1}	Mg C ha^{-1}	Country/ region	Source
African grass	—	—	8.67	—	Latin America	[50]
Andropogon guyanus		0–100	14.45	—	Latin America	[50, 51]
Brachiaria dictyoneura	3.5	—	8.57	30	Latin America	[52]
Lemongrass	—	0–30	3.08	—	India	[53]
"	—	0–30	5.38	—	India	[54]
Palmarosa	—	0–30	2.79	—	India	[53]
"	—	0–30	6.14	—	India	[54]
Kikuyu	3	0–10	2.6	34 g (kg^{-1})	Australia	[55]
"	15	0–20	—	67.2	Australia	[43]
"	—	0–30	0.9	—	West Australia	[56, 57]
"	—	0–30	0.26	—	South Australia	[56, 57]
Miscanthus	10	0–80	0.78	—	Europe	[58]
"	2.5	0–30	0.73	1.82	UK	[59]
"	2.5	0–30	0.87	2.17	UK	[59]
Vetiver	5	0–30	5.54	—	India	[53, 54]
"	7	0–30	1.61%		Ethiopia	[60]

Table 1.
Total soil carbon stored under different tropical grasses with different soil sampling depth and age of plantation.

Guo et al. [49], reported 15–20% larger soil C stocks under native pasture compared with a 16 year old pine plantation to 1.0 m in the soil profile. These findings seem to be convincing, although some caution must be attached to many such results given that they typically do not account for above-ground biomass and are rarely reported on an equivalent mass basis. There is nevertheless, growing evidence that tropical pastures might have the capacity to store SOC that is at least equivalent to that of forest systems in terms of rate and quantity of accumulation. However, the quantity and rate of carbon accumulation would appear to be moderated by environmental conditions and both preceding and ongoing management practices. Consideration and knowledge of the behavior and potential carbon storage of particular tropical grass species has much to add to this debate.

Some specific tropical grass species (**Table 1**) such as *Andropogon guyanus* (gamba grass, Rhodesian bluegrass, tambuki grass), Lemongrass (*Cymbopogon citratus*), Palmarosa (*Cymbopogon martinii*), Kikuyu (*Pennisetum clandestinum*); Miscanthus (*M. giganteus*), Vetiver (*Chrysopogon zizanioides*) have been highlighted for improving soil carbon storage potential even though their efficiency is determined by a range of environmental and management factors [46, 60–64]. However, Fearnside and Barbosa [62], found that management practices could on the other hand determine whether tropical pasture soils could be net sinks or sources of carbon, demonstrating in Brazilian Amazonia, that under "typical" (without inputs or other practices) and "ideal" (with variety of appropriate practices) management, tropical pasture soils were a net carbon source releasing an average of 12 Mg C ha^{-1} following deforestation.

3. Processes of SOC storage

The process by which organic carbon stored in soils follows various pathways such as roots, root exudates and litter (both above- and below-ground). Plant litter consists of dead roots, is a primary source of soil organic matter which is the largest terrestrial pool of carbon [65]. Despite, often considered separate processes of litter decomposition and soil organic matter stabilization is an important control of carbon storage and SOC dynamics [66, 67]. Decomposition of plant litter is one of the main processes driving nutrient and carbon (C) cycling in terrestrial ecosystems [68]. The effect of litter quality on SOM stabilization is inconsistent and litter addition promotes SOC mineralization, but this promotion alters by soil moisture and litter type [69]. Hence, understanding the interactions between the initial composition and subsequent decomposition of plant litter help to understand the flow of organic matter between soil carbon pools [70]. Root exudates are also one of the various pathways through which the carbons fixed released into soils [71]. Plants release a part of their metabolome into soils and thereby provide information about the potential biological function of exudates in the rhizosphere [72].

Root biomass production is an important plant component that can contribute to soil carbon sequestration. A strong fibrous root system, penetrating deep into the soil profile and growing vertically rather than horizontally, is therefore desirable to maximize soil carbon sequestration. Hence, the large root systems of tropical grasses might potentially facilitate long term deep carbon storage and reduce the chance of decomposition and carbon loss [44]. For example, the roots of vetiver grass have been found to contribute significantly more to additional SOC storage than those of other grass species [60, 63, 73]. Although the extent of SOC sequestration potential of tropical grass species still requires further research, they would appear to have particular promise with regard to soil carbon storage compared with other species.

Due to their large biomass production and their extensive and fast growing root system, tropical perennial grasses would seem to have the capacity to rapidly store or contribute large quantities of carbon in addition to their other varied uses [53, 74]. Deep rooted tropical perennial grasses have been identified as the most promising plants that could contribute to SOC storage and thus climate change mitigation [44, 75–78]. Awoke [79] highlighted further the potential of tropical grasses for both above- and below-ground C sequestration by planting strategically on appropriate lands.

Most of studies relating to tropical grasses to date have focused on the actual biomass production potential. However, there are only few studies which have considered the actual net accumulation of carbon stored in the soil under tropical grasses (**Table 1**) highlighting the need for controlled studies to determine not only biomass and inputs but also the net effect of tropical perennial grasses in terms of carbon storage and the mechanisms, stability and longevity of the carbon stored such as the rate of new carbon turnover and carbon cycling of the newly added carbon and the extent to which it is retained in the soil system.

4. Effect of cropland conversion to tropical pastures on soil carbon

Cropland conversion to pastures has recently become a common practice and is believed to have considerable potential to store significant quantities of additional SOC [19, 38, 80]. For example, Conant et al. [33] and Conant [81], reviewed studies worldwide and concluded that cropland conversion to grasslands can create a significant carbon sink, with a mean 5% annual increase in SOC. In the mid–western United States, agricultural land conversion to perennial grassland showed a constant rate of 62 g C m^{-2} $year^{-1}$ SOC accumulation over 40 years in the top 10 cm [82]. Similarly, Abberton et al. [83] reported that, in temperate regions, most grasslands can be considered soil carbon sinks of up to 40 g C m^{-2} $year^{-1}$ following cropland conversion. Post and Kwon [48], further estimated that land use change from cropping to grassland could result in an increase of 33.2 g soil C m^{-2} $year^{-1}$ in the USA. While a meta–analysis in temperate grasslands showed that at the 0–30 cm soil depth over 20 years SOC sequestration reached 44 g C m^{-2} $year^{-1}$ which is half of the rate (95 g C m^{-2} $year^{-1}$) at which SOC is lost over a 20 year period following permanent grassland conversion to an annual crop [84]. These estimates suggest that SOC recovery is possible but is usually slower than initial loss. Research in the south eastern United States also suggested up to 100 g C m^{-2} $year^{-1}$ could be sequestered in soil following conversion of cropland into optimally grazed pastures (where the available pasture matches the animal needs). These increases have been attributed to the fast growth habit of pastures, negligible erosion and the minimal disturbance to soil compared to cropping [28]. Although focused principally on temperate grass pasture species, these studies demonstrate the potential increase of SOC as a result of cropland conversion to grasslands.

5. Form and resilience of carbon stored under tropical grasses

Many of tropical pasture species have a distinctive carbon fixing (photosynthesis) pathway and are referred to as C_4 plants [43]. All plant species have the more primitive C_3 pathway, described by the Calvin Cycle [85] but an additional C_4 pathway evolved in species in the wet and dry tropics. C_4 pastures are those that have the photosynthetic processes divided between mesophyll and bundle sheath cells that are anatomically and biochemically separate, while C_3 pastures are those which use only the Calvin cycle photosynthesis pathway for fixing CO_2 which takes place inside of the chloroplast in mesophyll cells [86, 87].

In terms of photosynthetic efficiency, C_4 grasses are approximately 50% more efficient than C_3 plants as a result of this distinctive carbon fixation mechanism [88]. Wang et al. [87], indicated that more efficient use of light and CO_2 in C_4 plants results in an increase in both biomass production and CO_2 fixation. Hence, as a result of their high photosynthetic efficiency and productivity, tropical C_4 grasses might be expected to have larger potential for SOC sequestration compared with temperate and annual pastures [55]. Most tropical pastures are important perennials and provide a permanent soil cover and thus prevent soil surface erosion [89], which is of particular importance in the prevention of SOC loss by erosion. Greenland [90] hypothesized that, with suitable management practices, tropical grasses could have a significant potential as a soil carbon sink. Our knowledge of perennial tropical species growth, interaction with the soil, potential quantities and mechanisms of carbon storage remains incomplete [50].

It has been speculated that carbon storage in sub–soils might be an important mechanism leading to increased SOC storage in soils [44, 77] and it is known that tropical grasses translocate large quantities of carbon to their root systems [44]. This suggests an effective translocation to deeper soil layers where soil carbon is typically more protected from decomposition processes [91–94]. Accumulation of carbon in deeper soil layers might therefore be an important mechanism for carbon storage under this vegetation type [59, 89, 95]. The deep rootedness of tropical pastures might, therefore, potentially play an important role in transporting carbon to deeper soil layers and therefore facilitate SOC storage. Indeed, Fisher et al. [52], estimated that the introduction of deep rooted African grass pastures in Colombia might account for the sequestration of 100–507 Mt. soil carbon year^{-1} if their study sites were indeed representative of similar pastures throughout South America. These studies indicate the potential benefits of introducing deep rooted tropical perennial grasses for SOC storage but also the need for further carbon inventory.

6. Factors affecting SOC sequestration

Tropical pastures grow continually year round and are adapted to a wide range of soil and climate conditions because of the close interaction between climate factors and soil properties [28, 96]. In addition to soil type, management and site history could be important factors determining the direction and magnitude of change in soil carbon stock [28]. Similarly Chan and McCoy [43], indicated the higher effectiveness of pastures in increasing SOC storage under appropriate management. Wilson and Lonergan [97], also demonstrated in Australia that native and improved pastures in this environment had the same SOC quantity and that historical and contemporary management practice is a key factor influencing net SOC. The management of tropical pastures is therefore a critical determinant of whether the soils under this land use will represent a source or a sink of atmospheric carbon [62]. Poor pasture management such as over grazing, frequent burning and conversion to cultivated agricultural land could result in degradation and low productivity which can reverse the carbon sequestration potential of tropical pastures leading to carbon loss by erosion and oxidation [98]. Hence, the effects of tropical pastures on soil carbon are likely to vary because of environmental and management factors. For example Dalal et al. [99], demonstrated historical management as a key driver of SOC stock particularly in the surface soil layers. Therefore, there is a need for controlled studies that measure soil carbon with some certainty of the effects of both environmental and management factors.

Clay soils in general play a greater role to slow the rate of decomposition, due to both physical and chemical protection of SOC and typically promote larger soil carbon concentrations compared with sandy soils due to these SOC stabilization processes [100–102]. Similarly, a water saturated soil might have lower rates of organic matter breakdown because of a lack of oxygen for soil organisms compared to soils exposed to the atmosphere. Therefore, soil improvement and adding essential inputs are important to increase the rate of organic carbon addition and pasture production [28]. In addition, McKenzie and Mason [28], indicated that deep soil profiles with fertile subsoil allow deep root penetration into subsoil which is much cooler (less likely to promote decomposition) than the topsoil. Hence, maximizing the carbon input by increasing the net primary production through nutrient addition, increased nutrient and water use efficiency and minimizing the rate of organic matter decomposition after deposition in soil are important factors which can help to increase the amount of carbon sequestered from the atmosphere [96].

Carbon accumulation in pasture lands can also be determined by the length of time the land remains under pasture [64]. Hence, regardless of technologies or mechanisms, the length of time must also be taken into account when considering long–term carbon storage. Bouman et al. [103], stated that, due to various economic and biophysical dimensions, sustainability of tropical pastures can also be affected by the pasture type, age, and management which in turn can affect the carbon accumulation. Hence, McDermot and Elavarthi [104], recommended that best management practices, site specific policies and using technological options can offer good opportunities to generate positive effects on soil carbon accumulation by using tropical grasses.

Therefore, factors such as input versus outputs, climatic conditions, soil type and properties, land use control, management practices are the factors affecting SOC storage. Whenever there is a vegetation cover change from C_3 to C_4 plants, the ratio of stable carbon isotopes ($\delta^{13}C$) can be used to track changes in SOC between the C_3 and C_4 plants and the quantity of "new" carbon added [99, 105, 106]. The use of stable isotopes offers a useful quantitative technique to allow the estimation of organic carbon storage and turnover in soils, even when TOC changes are of limited magnitude [107].

7. Opportunities and economic benefits of tropical pastures

Biomass energy is currently receiving considerable attention in response to climate change and ever-increasing global energy demand [108–111] and tropical pastures would appear to have potential for the production of biofuels. For example, Clifton-Brown et al. [40], suggested that *Miscanthus* which is known for its high biomass production has value as a potential biofuel and the area over which it grows could therefore be expanded significantly throughout Europe. *Miscanthus* is grown in many European countries such as Austria, Denmark, France, Germany, Hungary, Poland, Switzerland and the United Kingdom. In France for instance, *Miscanthus* cultivation has increased since the first plantation in 2006 [112]. The EU Biofuel directive promotes the expansion of biofuels and *Miscanthus* particularly as a biomass energy source [113]. Hence, growing biomass energy crops (mainly tropical perennial grasses) specially in Europe is becoming common and expanding to consider for potential soil carbon storage and this needs to be explored further even in other parts of the world and using other different potential grasses.

8. Conclusion

Tropical pastures are potential candidates to contribute to climate change mitigation efforts through additional SOC storage due to their high biomass production, fast growth rates, and deep root systems. Using tropical grasses has also a low cost of implementation to rehabilitate degraded lands and improve soil productivity through increasing SOC. The existing literature on tropical grasses potential for soil carbon sequestration provides positive indications that significant soil carbon storage is possible. However, a number of further investigations are required to provide a sound basis on which management decisions involving tropical grasses can be made. We therefore recommend the following priority research actions to provide the required information:

- Cropland conversion to tropical perennial pastures and its potential to achieve multiple outcomes including soil health, soil security and sustainability in addition to soil carbon storage.

- A range of tropical grass species should be assessed to determine their potential to store additional carbon under specific climatic and management conditions.

- There is a need to assess the processes and mechanisms of SOC storage in deeper soil profiles under tropical perennial grasses to provide accurate estimates of SOC stocks.

- The rate of soil carbon turnover and cycling of the new carbon added and the extent to which it is retained in the soil system needs to be fully quantified.

- Best management practices, site specific policies and technological options which can positively affect soil carbon storage should be identified and clearly defined.

Author details

Bezaye Gorfu Tessema[1,2]*, Heiko Daniel[3], Zenebe Adimassu[4] and Brian Wilson[2,5]

1 Water and Land Resource Centre of Addis Ababa University, Addis Ababa, Ethiopia

2 School of Environmental and Rural Sciences, University of New England, Armidale, NSW, Australia

3 Research Services Directorate, University of New England, Armidale, NSW, Australia

4 International Water Management Institute, Accra, Ghana

5 NSW Department of Planning, Industry and Environment, Armidale, NSW, Australia

*Address all correspondence to: bezaye.g@wlrc-eth.org

References

[1] Fischer, G., Shah, M., & Van Velthuizen, H. (2002). Climate Change and Agricultural Vulnerability. *IIASA Publications Department*.

[2] Hamza, I., & Iyela, A. (2012). Land use pattern, climate change, and its implication for food security in Ethiopia: A review. Ethiopian Journal of Environmental Studies and Management, *5*(1), 26-31.

[3] Lal, R. (2004a). Soil carbon sequestration impacts on global climate change and food security. Science, *304*(5677), 1623-1627.

[4] Nelson, G. C., Rosegrant, M. W., Koo, J., Robertson, R., Sulser, T., Zhu, T., Ringler, C., Msangi, S., Palazzo, A., & Batka, M. (2009). *Climate change: Impact on agriculture and costs of adaptation* (Vol. 21): International Food Policy Research Institute, Washington, D.C.

[5] Rosenzweig, C., & Hillel, D. (1998). Climate change and the global harvest: potential impacts of the greenhouse effect on agriculture: Oxford University Press, UK.

[6] Girmay, G., & Singh, B. (2012). Changes in soil organic carbon stocks and soil quality: Land-use system effects in northern Ethiopia. Acta Agriculturae Scandinavica, Section B-Soil & Plant Science, *62*(6), 519-530.

[7] Lal, R. (2004b). Soil carbon sequestration to mitigate climate change. Geoderma, *123*(1), 1-22.

[8] Lambin, E. F., Turner, B. L., Geist, H. J., Agbola, S. B., Angelsen, A., Bruce, J. W., Coomes, O. T., Dirzo, R., Fischer, G., & Folke, C. (2001). The causes of land-use and land-cover change: Moving beyond the myths. Global environmental change - Human and Policy Dimensions, *11*(4), 261-269.

[9] Meshesha, D. T., Tsunekawa, A., Tsubo, M., Ali, S. A., & Haregeweyn, N. (2014). Land-use change and its socio-environmental impact in eastern Ethiopia's highland. Regional Environmental Change, *14*(2), 757-768.

[10] Chapin III, F. S., Carpenter, S. R., Kofinas, G. P., Folke, C., Abel, N., Clark, W. C., Olsson, P., Smith, D., Walker, B., & Young, O. R. (2010). Ecosystem stewardship: Sustainability strategies for a rapidly changing planet. Trends in Ecology & Evolution, *25*(4), 241-249.

[11] Lambin, E. F., Geist, H. J., & Lepers, E. (2003). Dynamics of land-use and land-cover change in tropical regions. Annual review of environment and resources, *28*(1), 205-241.

[12] Pielke Sr, R., Marland, G., Betts, R., Chase, T., Eastman, J., Niles, J., Niyogi, D., & Running, S. (2003). The influence of land-use change and landscape dynamics on the climate system: Relevance to climate-change policy beyond the radiative effect of greenhouse gases. Capturing carbon & conserving biodiversity–The market approach, 157-172.

[13] Eswaran, H., Van Den Berg, E., & Reich, P. (1993). Organic carbon in soils of the world. Soil Science Society of America Journal, *57*(1), 192-194.

[14] Jobbágy, E. G., & Jackson, R. B. (2000). The vertical distribution of soil organic carbon and its relation to climate and vegetation. Ecological Applications, *10*(2), 423-436.

[15] Batjes, N. (1998). Mitigation of atmospheric CO2 concentrations by increased carbon sequestration in the soil. Biology and Fertility of Soils, *27*(3), 230-235.

[16] Houghton, R. (2005a). Aboveground forest biomass and the

global carbon balance. Global Change Biology, *11*(6), 945-958.

[17] Houghton, R. (2005b). Tropical deforestation as a source of greenhouse gas emissions. Tropical deforestation and climate change, 13-21.

[18] Lal, R., & Follett, R. (2009). *Soil carbon sequestration and the greenhouse effect*: ASA-CSSA-SSSA.

[19] Freibauer, A., Rounsevell, M. D., Smith, P., & Verhagen, J. (2004). Carbon sequestration in the agricultural soils of Europe. Geoderma, *122*(1), 1-23.

[20] Lal, R. (2003). Global potential of soil carbon sequestration to mitigate the greenhouse effect. *Critical Reviews in Plant Sciences, 22*(2), 151-184. Doi:Doi 10.1080/713610854

[21] Lal, R., Griffin, M., Apt, J., Lave, L., & Morgan, M. G. (2004). Managing soil carbon. Science, *304*(5669), 393.

[22] Rabbi, S. M. F., Tighe, M., Cowie, A., Wilson, B. R., Schwenke, G., Mcleod, M., Badgery, W., & Baldock, J. (2014). The relationships between land uses, soil management practices, and soil carbon fractions in south eastern Australia. *Agriculture Ecosystems & Environment, 197*, 41-52. DOI:DOI 10.1016/j.agee.2014.06.020

[23] Smith, P., Bustamante, M., Ahammad, H., Clark, H., Dong, H., Elsiddig, E. A., Haber l, H., Harper, R., House, J., Jafari, M., Masera, O., Mbow, C., Ravindranath, N. H., Rice, C. W., Robledo Abad, C., Romanovskaya, A., Sperling, F., & Tubiello, F. (2014). Agriculture, Forestry and Other Land Use (AFOLU). In: Climate Change 2014: Mitigation of Climate Change. Contribution of Working Group III to the Fifth Assessment Report of the Intergovernmental Panel on Climate Change. *Cambridge University Press, Cambridge, United Kingdom and New York, NY, USA.*

[24] Rabbi, S. M. F., Hua, Q., Daniel, H., Lockwood, P. V., Wilson, B. R., & Young, I. M. (2013). Mean residence time of soil organic carbon in aggregates under contrasting land uses based on radiocarbon measurements. *Radiocarbon, 55*(1), 127-139. Retrieved from <go to ISI>://WOS:000315338000011

[25] Gallo, M. E., Porras-Alfaro, A., Odenbach, K. J., & Sinsabaugh, R. L. (2009). Photoacceleration of plant litter decomposition in an arid environment. Soil Biology and Biochemistry, *41*(7), 1433-1441.

[26] Grandy, S. A., & Robertson, P. G. (2006). Initial cultivation of a temperate-region soil immediately accelerates aggregate turnover and CO_2 and N_2O fluxes. Global Change Biology, *12*(8), 1507-1520.

[27] Janik, L. J., Skjemstad, J., Shepherd, K., & Spouncer, L. (2007). The prediction of soil carbon fractions using mid-infrared-partial least square analysis. Soil Research, *45*(2), 73-81.

[28] McKenzie, D., & Mason, W. (2010). Soil carbon sequestration under pasture in southern Australia. *Inquiry into soil sequestration in Victoria, Submission no29, Prepared for Dairy Australia, Project MCK13538*.

[29] Schuman, G., Ingram, L., Stahl, P., Derner, J., Vance, G., & Morgan, J. (2009). Influence of management on soil organic carbon dynamics in northern mixed-grass rangeland. *Soil carbon sequestration and the greenhouse effect, Second edition.(eds. R Lal, R Follett)*, 169-180.

[30] Lemma, B., Kleja, D. B., Olsson, M., & Nilsson, I. (2007). Factors controlling soil organic carbon sequestration under exotic tree plantations: A case study using the CO2Fix model in southwestern Ethiopia. Forest ecology and management, *252*(1), 124-131.

[31] Shiferaw, A., Hurni, H., & Zeleke, G. (2013). A review on soil carbon sequestration in Ethiopia to mitigate land degradation and climate change. Journal of Environment and Earth Science, *3*(12), 187-200.

[32] Oladele, O. I., & Braimoh, A. K. (2011). Soil carbon for food security and climate change mitigation and adaptation. Italian Journal of Agronomy, *6*(4), 38.

[33] Conant, R., Paustian, K., & Elliott, E. T. (2001). Grassland management and conversion into grassland: Effects on soil carbon. Ecological Applications, *11*(2), 343-355.

[34] Davidson, E. A., & Ackerman, I. L. (1993). Changes in soil carbon inventories following cultivation of previously untilled soils. Biogeochemistry, *20*(3), 161-193.

[35] Paustian, K., Collins, H. P., & Paul, E. A. (1997). *Management controls on soil carbon*: 1997a, Chemical Rubber Company (CRC) Press: Boca Raton, FL, USA.

[36] Young, R., Wilson, B. R., McLeod, M., & Alston, C. (2005). Carbon storage in the soils and vegetation of contrasting land uses in northern New South Wales, Australia. Soil Research, *43*(1), 21-31.

[37] Lesslie, R., Hill, M. J., Woldendorp, G., Dawson, S., & Smith, J. (2006). *Towards Sustainability for Australia's Rangelands: Analysing the Options*: Bureau of Rural Sciences.

[38] Schuman, G., Janzen, H. H., & Herrick, J. E. (2002). Soil carbon dynamics and potential carbon sequestration by rangelands. Environmental Pollution, *116*(3), 391-396.

[39] Janssens, I., Freibauer, A., Schlamadinger, B., Ceulemans, R., Ciais, P., Dolman, A., Heimann, M., Nabuurs, G.-J., Smith, P., & Valentini, R. (2005). The carbon budget of terrestrial ecosystems at country-scale–a European case study. Biogeosciences, *2*(1), 15-26.

[40] Clifton-Brown, J. C., Breuer, J., & Jones, M. B. (2007). Carbon mitigation by the energy crop, Miscanthus. Global Change Biology, *13*(11), 2296-2307.

[41] Schwenke, G., McLeod, M., Murphy, S., Harden, S., Cowie, A., & Lonergan, V. (2014). The potential for sown tropical perennial grass pastures to improve soil organic carbon in the north-west slopes and plains of New South Wales. Soil Research, *51*(8), 726-737.

[42] Parton, W., Scurlock, J., Ojima, D., Gilmanov, T., Scholes, R., Schimel, D. S., Kirchner, T., Menaut, J. C., Seastedt, T., & Garcia Moya, E. (1993). Observations and modeling of biomass and soil organic matter dynamics for the grassland biome worldwide. Global Biogeochemical Cycles, *7*(4), 785-809.

[43] Chan, K., & McCoy, D. (2010). Soil carbon storage potential under perennial pastures in the mid-north coast of New South Wales, Australia. Tropical Grasslands, *44*, 184-191.

[44] Lavania, U., & Lavania, S. (2009). Sequestration of atmospheric carbon into subsoil horizons through deep-rooted grasses-vetiver grass model. Current Science, *97*(5), 618-619.

[45] Tubiello, F. N., Soussana, J.-F., & Howden, S. M. (2007). Crop and pasture response to climate change. Proceedings of the National Academy of Sciences, *104*(50), 19686-19690.

[46] Amézquita, M., Murgueitio, E., Ibrahim, M., & Ramírez, B. (2010). Carbon sequestration in pasture and silvopastoral systems compared with native forests in ecosystems of tropical America. Grassland carbon

sequestration: management, policy and economics, *11*, 153-161.

[47] Desjardins, T., Barros, E., Sarrazin, M., Girardin, C., & Mariotti, A. (2004). Effects of forest conversion to pasture on soil carbon content and dynamics in Brazilian Amazonia. Agriculture, Ecosystems & Environment, *103*(2), 365-373.

[48] Post, W. M., & Kwon, K. C. (2000). Soil carbon sequestration and land-use change: Processes and potential. Global change biology, *6*(3), 317-327.

[49] Guo, L., Cowie, A. L., Montagu, K. D., & Gifford, R. M. (2008). Carbon and nitrogen stocks in a native pasture and an adjacent 16-year-old Pinus radiata D. Don plantation in Australia. Agriculture, Ecosystems & Environment, *124*(3), 205-218.

[50] Fisher, M. J., Braz, S., Dos Santos, R., Urquiaga, S., Alves, B., & Boddey, R. (2007). Another dimension to grazing systems: Soil carbon. Tropical Grasslands, *41*(2), 65.

[51] Mishra, V., Ranade, D., Joshi, R., & Sharma, R. (1997). Vetiver: A miracle grass. Review of an international conference. Crop Research (Hisar), 13(2), 507-509.

[52] Fisher, M. J., Rao, I. M., Ayarza, M. A., Lascano, C. E., Sanz, J., Thomas, R. J., & Vera, R. R. (1994). Carbon storage by introduced deep-rooted grasses in the South American savannas. *371.*

[53] Singh, M., Guleria, N., Rao, E. V. P., & Goswami, P. (2013). Efficient C sequestration and benefits of medicinal vetiver cropping in tropical regions. Agronomy for Sustainable Development, 1-5.

[54] Singh, M., Guleria, N., Rao, E. P., & Goswami, P. (2011). A Strategy for Sustainable Carbon Sequestration using Vetiver (Vetiveria zizanioides (L.)): A Quantitative Assessment over India. Project Document CM PD-1101.

[55] Neal, J., Eldridge, S., Fulkerson, W., Lawrie, R., & Barchia, I. (2013). Differences in soil carbon sequestration and soil nitrogen among forages used by the dairy industry. Soil Biology and Biochemistry, *57*, 542-548.

[56] Sanderman, J., Fillery, I., Jongepier, R., Massalsky, A., Roper, M., Macdonald, L., Maddern, T., Murphy, D., & Baldock, J. (2013a). Carbon sequestration under subtropical perennial pastures II: Carbon dynamics. Soil Research, 51(8), 771-780.

[57] Sanderman, J., Fillery, I., Jongepier, R., Massalsky, A., Roper, M., Macdonald, L., Maddern, T., Murphy, D., Wilson, B., & Baldock, J. (2013b). Carbon sequestration under subtropical perennial pastures I: Overall trends. Soil Research, 51(8), 760-770.

[58] Poeplau, C., & Don, A. (2013). Soil carbon changes under Miscanthus driven by C4 accumulation and C3 decomposition–toward a default sequestration function. Global Change Biology Bioenergy, 6(4), 327-338.

[59] Zimmermann, J., Dauber, J., & Jones, M. B. (2012). Soil carbon sequestration during the establishment phase of Miscanthus× giganteus: A regional-scale study on commercial farms using 13C natural abundance. Global Change Biology Bioenergy, *4*(4), 453-461.

[60] Hailu, L., Tesfaye, G., & Yaekob, T. (2020). Effect of Vetiver Grass (Vetiver Zizanodes) Hedgerows on Selected Soil Properties and Crop Yield on Farm Land at Haru District, Western Ethiopia. *International Journal of Research Studies in Agricultural Sciences, 6*(5). doi:10.20431/2454-6224.0605005

[61] Amézquita, M., Ibrahim, M., Llanderal, T., Buurman, P., &

Amézquita, E. (2004). Carbon sequestration in pastures, silvo-pastoral systems and forests in four regions of the Latin American tropics. Journal of Sustainable Forestry, *21*(1), 31-49.

[62] Fearnside, P. M., & Barbosa, R. I. (1998). Soil carbon changes from conversion of forest to pasture in Brazilian Amazonia. Forest ecology and management, *108*(1-2), 147-166.

[63] Lakshmi, C. S., & Sekhar, C. C. (2020). Role of *Vetiveria zizanioides* in Soil Protection and Carbon Sequestration.

[64] Neill, C., Melillo, J. M., Steudler, P. A., Cerri, C. C., de Moraes, J. F., Piccolo, M. C., & Brito, M. (1997). Soil carbon and nitrogen stocks following forest clearing for pasture in the southwestern Brazilian Amazon. Ecological Applications, *7*(4), 1216-1225.

[65] Schmidt, M. W., Torn, M. S., Abiven, S., Dittmar, T., Guggenberger, G., Janssens, I. A., Kleber, M., Kogel-Knabner, I., Lehmann, J., Manning, D. A., Nannipieri, P., Rasse, D. P., Weiner, S., & Trumbore, S. E. (2011). Persistence of soil organic matter as an ecosystem property. *Nature, 478*(7367), 49-56. Retrieved from http://www.ncbi.nlm.nih.gov/pubmed/21979045

[66] Castellano, M. J., Mueller, K. E., Olk, D. C., Sawyer, J. E., & Six, J. (2015). Integrating plant litter quality, soil organic matter stabilization, and the carbon saturation concept. Glob Chang Biol, *21*(9), 3200-3209. doi:10.1111/gcb.12982

[67] Don, A., & Kalbitz, K. (2005). Amounts and degradability of dissolved organic carbon from foliar litter at different decomposition stages. Soil Biology and Biochemistry, *37*(12), 2171-2179.

[68] Solly, E. F., Schöning, I., Boch, S., Kandeler, E., Marhan, S., Michalzik, B., Müller, J., Zscheischler, J., Trumbore, S. E., & Schrumpf, M. (2014). Factors controlling decomposition rates of fine root litter in temperate forests and grasslands. Plant and Soil, *382*(1-2), 203-218. doi:10.1007/s11104-014-2151-4

[69] Wang, Q., Zeng, Z., & Zhong, M. (2016). Soil moisture alters the response of soil organic carbon mineralization to litter addition. Ecosystems.

[70] Walela, C., Daniel, H., Wilson, B., Lockwood, P., Cowie, A., & Harden, S. (2014). The initial lignin: Nitrogen ratio of litter from above and below ground sources strongly and negatively influenced decay rates of slowly decomposing litter carbon pools. *Soil Biology and Biochemistry, 77*, 268-275. Retrieved from <go to ISI>://WOS:000341556600030

[71] Guo, L., Halliday, M., Siakimotu, S., & Gifford, R. (2005). Fine root production and litter input: Its effects on soil carbon. Plant and Soil, *272*(1-2), 1-10.

[72] Dietz, S., Herz, K., Gorzolka, K., Jandt, U., Bruelheide, H., & Scheel, D. (2020). Root exudate composition of grass and forb species in natural grasslands. Scientific reports, *10*(1), 1-15.

[73] Grimshaw, R. G. (2008). The Vetiver System. First National Indian Vetiver Workshop (Pp. 21-23). Kochi, India: The Vetiver International.

[74] Mondyagu, S., Kopsell, D. E., Steffen, R. W., Kopsell, D. A., & Rhykerd, R. L. (2012). The effect of nitrogen level and form on the growth and development of vetiver grass (Chrysopogon zizanioides). Transactions of the Illinois State Academy of Science, *105*, 1-10.

[75] Abate, H., & Simane, B. (2010). Multiple Benefits of the Vetiver System

and its Environmental Application in Ethiopia. *Feature Article*.

[76] Gaspard, S., Altenor, S., Dawson, E. A., Barnes, P. A., & Ouensanga, A. (2007). Activated carbon from vetiver roots: Gas and liquid adsorption studies. Journal of Hazard Mater, *144*(1-2), 73-81.

[77] Sida, S. (2010). Vetiver system applications for enhancing ecological and social systems. Strengthening Capacity for Climate Change Adaptation in the Agriculture Sector in Ethiopia, 81-88.

[78] Truong, P. (2000). *The Global Impact of Vetiver Grass Technology on the Environment.* Paper presented at the Proceedings of the second international Vetiver conference, Thailand.

[79] Awoke, T. C. (2013). Conservation Markating: The case of vetiver grass technology (VGT) in Illu aba bora zone. Journal of Marketing and Consumer Research, *2*, 1-7.

[80] Derner, J., & Schuman, G. (2007). Carbon sequestration and rangelands: A synthesis of land management and precipitation effects. Journal of Soil and Water Conservation, *62*(2), 77-85.

[81] Conant, R. (2012). Grassland soil organic carbon stocks: Status, opportunities, Vulnerability. 275-302.

[82] McLauchlan, K., Hobbie, S., & Post, W. (2006). Conversion from agriculture to grassland builds soil organic matter on decadal timescales. Ecological Applications, *16*(1), 143-153.

[83] Abberton, M., Conant, R., & Batello, C. (2010). Grassland carbon sequestration: Management, policy and economics. Integrated Crop Management, *11*, 1-18.

[84] Soussana, J. F., Loiseau, P., Vuichard, N., Ceschia, E., Balesdent, J., Chevallier, T., & Arrouays, D. (2004). Carbon cycling and sequestration opportunities in temperate grasslands. Soil use and management, *20*(2), 219-230.

[85] Ludlow, M., Fisher, M., & Wilson, J. (1985). Stomatal adjustment to water deficits in three tropical grasses and a tropical legume grown in controlled conditions and in the field. Functional Plant Biology, *12*(2), 131-149.

[86] Ludlow, M. (1985). Photosynthesis and dry matter production in C3 and C4 pasture plants, with special emphasis on tropical C3 legumes and C4 grasses. Functional Plant Biology, *12*(6), 557-572.

[87] Wang, C., Guo, L., Li, Y., & Wang, Z. (2012). Systematic comparison of C3 and C4 plants based on metabolic network analysis. *BioMed Central Systems Biology, 6 Suppl 2*, S9. Retrieved from http://www.ncbi.nlm.nih.gov/pubmed/23281598

[88] Kajala, K., Covshoff, S., Karki, S., Woodfield, H., Tolley, B. J., Dionora, M. J. A., Mogul, R. T., Mabilangan, A. E., Danila, F. R., & Hibberd, J. M. (2011). Strategies for engineering a two-celled C4 photosynthetic pathway into rice. Journal of Experimental Botany, *62*(9), 3001-3010.

[89] Peters, M., Rao, I., Fisher, M., Subbarao, G., Martens, S., Herrero, M., van der Hoek, R., Schultze-Kraft, R., Miles, J., & Castro, A. (2012). Tropical forage-based systems to mitigate greenhouse gas emissions. 171-190.

[90] Greenfield, J. (1988). Vetiver Grass, A Little Recent History and Thoughts for the Future. (Advance of Scientific Research on Vetiver System), 500-506.

[91] Fontaine, S., Barot, S., Barré, P., Bdioui, N., Mary, B., & Rumpel, C. (2007). Stability of organic carbon in deep soil layers controlled by fresh

carbon supply. Nature, *450*(7167), 277-280.

[92] Harrison, R. B., Footen, P. W., & Strahm, B. D. . (2011). Deep soil horizons: Contribution and importance to soil carbon pools and in assessing whole-ecosystem response to management and global change. Forest Science, *57*(1), 67-76.

[93] Mathieu, J. A., Hatté, C., Balesdent, J., & Parent, É. (2015). Deep soil carbon dynamics are driven more by soil type than by climate: A worldwide meta-analysis of radiocarbon profiles. Global Change Biology, *21*(11), 4278-4292.

[94] Rumpel, C., & Kögel-Knabner, I. (2011). Deep soil organic matter—A key but poorly understood component of terrestrial C cycle. Plant and Soil, *338* (1), 143-158.

[95] Kuzyakov, Y., & Domanski, G. (2000). Carbon input by plants into the soil. Review. *Journal of Plant Nutrition and Soil Science, 163*(4), 421-431.

[96] Reichle, D., Houghton, J., Kane, B., Ekmann, J., Benson, S., Clarke, J., Dahlman, R., Hendrey, G., Herzog, H., & Huntercevera, J. (1999). Carbon Sequestration: State of the Science: US Department of Energy. *Offices of Science and Fossil Energy, USA*.

[97] Wilson, B. R., & Lonergan, V. E. (2013). Land-use and historical management effects on soil organic carbon in grazing systems on the northern tablelands of New South Wales. Soil Research, *51*(8), 668.

[98] Scurlock, J., & Hall, D. (1998). The global carbon sink: A grassland perspective. Global Change Biology, *4*(2), 229-233.

[99] Dalal, R. C., Thornton, C. M., & Cowie, B. A. (2013). Turnover of organic carbon and nitrogen in soil assessed from δ 13 C and δ 15 N changes under pasture and cropping practices and estimates of greenhouse gas emissions. Science of the Total Environment, *465*, 26-35.

[100] Kirschbaum, M. U., Moinet, G. Y., Hedley, C. B., Beare, M. H., & McNally, S. R. (2020). A conceptual model of carbon stabilisation based on patterns observed in different soils. Soil Biology and Biochemistry, *141*, 107683.

[101] Rasse, D. P., Rumpel, C., & Dignac, M. F. (2005). Is soil carbon mostly root carbon? Mechanisms for a specific stabilisation. Plant and Soil, *269*(1), 341-356.

[102] Saidy, A. R., Smernik, R. J., Baldock, J. A., Kaiser, K., Sanderman, J., & Macdonald, L. M. (2012). Effects of clay mineralogy and hydrous iron oxides on labile organic carbon stabilisation. Geoderma, *173*, 104-110.

[103] Bouman, B. A. M., Plant, R. A. J., & Nieuwenhuyse, A. (1999). Quantifying economic and biophysical sustainability trade-offs in tropical pastures. Ecological Modelling, *120*(1), 31-46.

[104] McDermot, C., & Elavarthi, S. (2014). Rangelands as carbon sinks to mitigate climate change: A review. Journal of Earth Science & Climatic Change, *05*(08).

[105] Ehleringer, J. R., Buchmann, N., & Flanagan, L. B. (2000). Carbon isotope ratios in belowground carbon cycle processes. Ecological Applications, *10*(2), 412-422.

[106] Schneckenberger, K., & Kuzyakov, Y. (2007). Carbon sequestration under Miscanthus in sandy and loamy soils estimated by natural 13C abundance. Journal of Plant Nutrition and Soil Science, *170*(4), 538-542.

[107] Dalal, R. C., Cowie, B., Allen, D., & Yo, S. (2011). Assessing carbon

lability of particulate organic matter from δ13C changes following land-use change from C3 native vegetation to C4 pasture. Soil Research, *49*(1), 98-103.

[108] Berndes, G., & Hansson, J. (2007). Bioenergy expansion in the EU: Cost-effective climate change mitigation, employment creation and reduced dependency on imported fuels. Energy policy, *35*(12), 5965-5979.

[109] de Wit, M. P., Lesschen, J. P., Londo, M. H., & Faaij, A. P. (2014). Greenhouse gas mitigation effects of integrating biomass production into European agriculture. Biofuels, Bioproducts and Biorefining, *8*(3), 374-390.

[110] Hoogwijk, M., Faaij, A., Eickhout, B., de Vries, B., & Turkenburg, W. (2005). Potential of biomass energy out to 2100, for four IPCC SRES land-use scenarios. Biomass and Bioenergy, *29*(4), 225-257.

[111] Sims, R. E., Hastings, A., Schlamadinger, B., Taylor, G., & Smith, P. (2006). Energy crops: Current status and future prospects. Global Change Biology, *12*(11), 2054-2076.

[112] NovaBiom. (2015). Miscanthus : A Profitable Crop, Thriving in Europe Since The 2000's. *http://www.novabiom.com/en/miscanthuseng, Accessed on 12/62015*.

[113] Edwards, R., Mulligan, D., & Marelli, L. (2010). Indirect land use change from increased biofuels demand: Comparison of models and results for marginal biofuels production from different feedstocks., EC Joint Research Centre, Ispra. *Joint Research Centre Scientific and Technical Reports*.

Chapter 3

Optimization and Characterization of Novel and Non-Edible Seed Oil Sources for Biodiesel Production

Inam Ullah Khan and Syed Aftab Hussain Shah

Abstract

Biodiesel mainly comes from edible oil, and there is little research on its yield from non-edible sources with low-cost oil. It is paramount to investigate the non-edible oil resources which may lead to advance the commercial feasibility of biodiesel and cost effectiveness as well as resolve the food issues. This chapter describes four novel non-edible seed oil sources comprising *Koelreuteria paniculata, Rhus typhina, Acacia farnesiana* and *Albizzia julibrissin* for biodiesel production. We aimed to optimize different reaction parameters for oil extraction, alkali-catalyzed transesterification process for maximal biodiesel production and finally evaluate its compatibility with mineral diesel. The optimization factors in transesterification included the molar ratio of methanol to oil, reaction time, stirring intensity, catalyst concentration and temperature. Two methods have been described including Soxhlet and mechanical for extraction of seed oil. The synthesized esters were evaluated and characterized through the nuclear magnetic resonance (NMR; ^{1}H and ^{13}C), Fourier transform infrared (FT-IR) and gas chromatography–mass spectrometry (*GC–MS*) and the total conversion of crude oil to fatty acid methyl esters (FAMEs) were established. The inductively coupled plasma-optical emission spectrometry (ICP-OES) and Elemental Analyzer (EA) were used for evaluation of elemental concentration. The physico-chemical characterizations of the biodiesel, i.e., flash point, pour point, cloud point, and density were within the *American Society for Testing and Materials* (ASTM; D6751) and European Standards ((EN14214). *Koelreuteria paniculata* produced highest biodiesel oil content by Soxhlet extraction (28–30%) followed by the *Albizzia julibrissin* (19–24%), *Acacia farnesiana* (23%), *Rhus typhina* (20–22%). The density ranged from 0.83–0.87 @ 15°C (g/cm^3) and the kinematic viscosity ranged from 3.75–6.3 (mm^2/s) among all the plant sources. *Koelreuteria paniculata* had highest Na (5456.2), Cr (1246.8), Ni (658.36), and Al (346.87) elemental concentrations (μg/g) than other plant sources. The elemental percent of C, H, N, and O of biodiesel ranged from 72.54–76.86, 11.25–13.34, 1.97–2.73, and 9.86–12, respectively. In conclusion, these non-edible plant seeds offer a cheap source of renewable energy and can be easily grown on barren and wastelands and contribute to efficient biodiesel production to mitigate the energy crisis.

Keywords: *Acacia farnesiana*, *Albizzia julibrissin*, *Rhus typhina*, *Koelreuteria paniculata*, Biodiesel

1. Introduction

The global energy need has been confronting major challenges owing to population growth and industrialization [1, 2]. Green house gases and their emissions as well as developing energy safety mechanisms have perpetually turned the focus on research and technological development in this sector. The researcher community is applying renewable energy practices as an alternate to petroleum fuels with biodiesel, bioethanol, biomass, biogas, and synthetic fuels with the aim to curtail net CO_2 emission, and improve air, soil, water and global warming [3]. The American Biodiesel Standard Specification (ASTM 6751) defines biodiesel (also named fatty acid methyl ester; FAMEs) as fuel comprising of monoalkyl esters of long-chain fatty acids acquired from vegetable oils or animal fats [4]. The International Energy Agency (IEA) provided the estimates about global market share of biofuels to be increased from 1% (2004) to 7% by 2030 [5]. The need for utilizing biodiesel is associated with its lower exhaust emissions (COx, SOx) and particulate matter [6]. Moreover, it possesses tremendous biodegradability [7], lubricity, storage [8], and higher flash point [6], oxygen content than diesel [9–11]. The higher oxygen content reflects the low carbon emissions, particulate emissions, CO, aromatic hydrocarbons, sulfur, smoke, and noise [12]. The major issues for biodiesel production and commercialization from vegetable oils comprise their availability and manufacturing cost [13].

The raw materials of biodiesel can be classified into three major groups including vegetable oil (edible or non-edible oil), animal fat, and edible waste oil [14]. These sources possess triglycerides [15] which carry great potential. Biodiesel obtained from vegetable oil has a viable market share in USA and European countries [16]. The scientific community is facing eminent challenge remains for suitable raw materials, their extraction and finally characterization for efficient and cost-effective biodiesel production. The transesterification [17] is a specialized method for biodiesel production from vegetable sources through conversion of one ester to another having low viscosity than the mineral diesel. The transesterification reaction involves catalyst between triglycerides, and short-chain alcohols, which produce monoesters, branched-chain, and long-chain triglyceride molecules that are further converted into glycerol and monoesters [18]. The three-step reaction forms monoglycerides and diglycerides as intermediates. As methanol contains lower charge, it is efficiently used for commercial production of biodiesel. Potassium hydroxide (KOH) as a predominate role in transesterification reaction [19]. The palm, sunflower, coconut rapeseed, soybean, and flaxseed are some of the raw materials being employed for commercialization [20]. Vegetable oil contains complex structure so it cannot be directly used in diesel engines and it will further aggravate the food supply chain through depletion of forests and wildlife destruction. Thus, impetus, toward non-edible sources, has been shifted for biodiesel production.

Feedstock has greater significance for ample availability of biodiesel [21]. The redeeming traits of non-edible sources include their toxicity, no utility in human food as it contains Erucic acid as major constituent of fatty acid; 56–66% [22], and its easy cultivation on poor soils [23], and cost-effectiveness. Moreover, it is very stable and possesses low melting point [24]. Biomass is a major energy source covering almost 10–14% of global need due to its easy combustion, less pollution and lower ash content [8]. However, it has equally low calorific value, thermal efficiency (10 to 15%), and comparatively large volume and transportation issues [8]. Chemically, biomass energy can be converted into liquid and gaseous forms [25].

Many studies have been conducted to explore the non-edible sources for biodiesel production comprising *Croton megalocarpus* [26], *Prunus dulcis* [27], *Prunus sibirica* [28], *Rhazya stricta* Decne [29], rubber seed oil [30], *Silybum marianum* L. [31], wild *Brassica juncea* L. [32], *Jatropha curcas* and Karanja [33–35], waste tallow [36], and notably, algae [37]. However, high-quality biodiesel production still remains to decipher from existing economical non-edible sources [38].

2. Description of non-edible plant sources

Four novel and non-edible plant sources have been explored for bio-diesel production comprising *Koelreuteria paniculata* [39], *Rhus typhina* [40], *Acacia farnesiana* [41], and *Albizzia julibrissin* [42]. The plant seeds were collected from China (Urumqi, Binhai new area near Nankai University's new campus Tianjin) and Pakistan (Lakki Marwat and Islamabad).

2.1 *Koelreuteria paniculata*

Koelreuteria paniculata belongs to the family Sapindaceae. It is a novel non-edible seed oil source that can be investigated for biodiesel production (**Figure 1**). It is less likely that pests may destroy it. It can grow at different soil

Figure 1.
Koelreuteria paniculata. (A) Plant; (B) seeds; (C) biodiesel; (D) glycerin.

environments, even at high pH soils. It possesses gorgeous inflorescence with ellipsoid pods and contains abundance of seeds. About 15–20kg seeds are produced annually from a single plant. The feasibility of planting *K. paniculata* trees ranges from 400 to 500 per hectare area with average production of 115,000 kg seeds and 30,000 kg per hectare oil production. *K. paniculata* species have origin from China, Japan, and Korea. *K. paniculata* is often used as an ornamental plant and declared as an invasive due to the inexhaustible seed production and offspring. *K. paniculata* is suitable to grow in unused and desolate lands and its seeds can be effectively used as alternate energy source, which contain 28–30% oil constituents [39, 43].

2.2 *Rhus typhina* L.

Rhus typhina L. (Staghorn sumac) is belongs to Anacardiaceae family (**Figure 2**). It has greater distribution in subtropical and temperate regions around the world, notably in Africa and North America. It has a maximum height ranging from 30 to 35 feet. Its seeds contain enormous potential as feedstock for biodiesel production [40]. *R. typhina* plant grows rapidly and produces abundance of seeds. About 300–400 trees can be planted, with approximate 78,000 kg seeds

Figure 2.
Rhus typhina (staghorn sumac). (A) Plant; (B) seeds; (C) biodiesel; (D) glycerin.

and 17,160 kg production per hectare, respectively. *R. typhina* oil yield has been reported as 9% w/w on the basis of acetone/water extract 9:1 v/v [44] and 12% [45]; wherein we obtained 22% oil contents [40].

2.3 *Acacia farnesiana* L.

Acacia farnesiana L. belongs to family Leguminosae (Mimosoideae) and it is native to North America [46]. This tall semi-evergreen native shrub or small tree is commonly referred to as sweet acacia, Huisache, etc., with soft, medium-green feather-like, finely divided small leaves. The slightly thick stem is rich in chocolate brown or gray, with long and pointed needles. The small, puff-like yellow flowers are very fragrant, appear in clusters in late winter, and then occasionally spread out after each new flush, providing nearly four seasons of flowering. An area of about one hectare wills 91,500 kg of seeds yield, and the efficiency of oil per hectare is approximately 21,250 kg. The fruit is an elongated pod, 3 to 6 inches long, dry, and covered with hard skin, brown. Green color attracts birds; squirrels and other mammals have no obvious littering problems and stick to the trees, which is very beautiful. The long-lasting fruit has a smooth appearance and contains seeds cherished by birds and other wildlife (**Figure 3**).

Figure 3.
Acacia farnesiana. (A) Plant; (B) seeds; (C) biodiesel; (D) glycerin.

Figure 4.
Albizzia julibrissin. (A) Plant; (B) seeds; (C) biodiesel; (D) glycerin.

2.4 *Albizzia julibrissin*

One of the plants, *Albizzia julibrissin* (Mimosa) belong to Fabaceae, which contains non-edible seeds, can be evaluated for biodiesel production (**Figure 4**). *A. julibrissin* is a deciduous tree, wide and 12 m (40ft) high, but usually 3–6 m (10–20ft), smooth bark, gray. *A. julibrissin* fruit comprises of lentil pods with enlarged seeds. Each pod is 8–18 cm long and 1.5–2.5 cm wide and can be observed from June to February. Each pod usually grows 5–10 elliptical seeds, about 1.25 in length. Some authors assert that mimosa produces many seeds [47]. *A. julibrissin* can produce 8,000 seeds per year. The average *A. julibrissin* seed per pound is 11,000 to 11,500 [48, 49]. Wind, gravity, and water are major contributory factors for dispersal of seeds and pods. *A. julibrissin* is native to Asia and found in Turkey, Azerbaijan, China, Japan, Taiwan, and other temperate regions, Bhutan in Asia, India, Nepal, Pakistan, Myanmar, Japan and other tropical areas [42, 50].

3. Oil extraction methodologies and outcomes from non-edible plant sources

3.1 Oil extraction

After shelling, drying and pulverizing with a grinder (Xiantaopai XTP-10000A, Zhejiang, China) the seeds were processed for extraction. The oil extraction from

Acacia farnesiana, Albizzia julibrissin, Rhus typhina and *Koelreuteria paniculata* was carried by Soxhlet (chemical; 90°C for 7 h using different solvents; **Figure 5A**) and mechanical extraction of by two different electric oil expeller machines comprising Fangtai Shibayoufang FL-S2017 China (less power extractor) and Fangtai Shibayoufang J508, China (high power extractor), (**Figure 5B**). Pre-treatment of seed is essential for mechanical and Soxhlet extraction, which can increase the amount of oil recovery. After 2–3 revolutions, a considerable yield of crude seed oil was obtained. The oil removed from the seed by mechanical presses requires additional handling of extraction and filtration to produce a purer raw feedstock. Further following steps were conducted to get the biodiesel from these sources comprising filtration, rotary evaporation for access methanol, heating, transesterification, settling, separation, and washing.

The oil production was calculated by the following equation.

$$\text{Conversion\%} = \text{Obtained seed oil weight}\,(\text{g}) / \text{Total seed weight}\,(\text{g}) \times 100 \quad (1)$$

The comparative oil content (%) of 4 plants, obtained by using mechanical oil extraction and Soxhlet extraction methods from four plant sources is given in **Table 1**.

3.2 Fourier-transform infrared spectroscopy (FT-IR)

FT-IR spectroscopy data of the mid-infrared region of biodiesel samples to recognize functional groups and the bands analogous to various stretching and bending vibrations is highlighted in **Table 2**.

3.3 Nuclear magnetic resonance (NMR)

The FAMEs NMR spectrum was acquired by (Bruker Avance III 400 NMR Spectrometer, Karlsruhe, Germany) at 400 MHz (^{1}H-NMR) or 100 MHz (^{13}C-NMR). Denatured chloroform was used as solvent and tetramethylsilane as the internal standard. The biodiesel ^{1}H NMR (300 MHz) spectrum was noted with a cycle delay of 1.0 s, and eight times scans with a pulse duration of 30°, (**Table 3**). A carbon ^{13}C NMR (75 MHz) spectrum was recorded with pulse duration of 30°and a cycle delay of 1.89 s, followed by scanning for 160 times (**Table 4**).

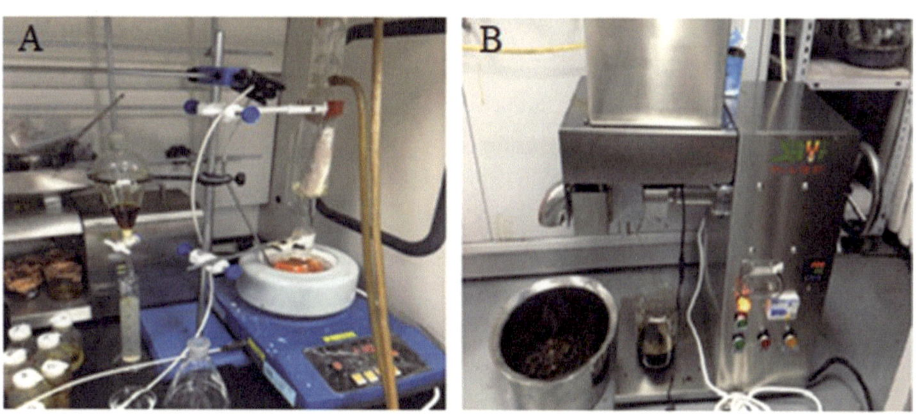

Figure 5.
(A) Instrument for Soxhlet extraction (chemical extraction); (B) instrument for mechanical oil extraction.

S. No.	Source Name	Mechanical Extraction (%)	Soxhlet extraction (%)	FFAs content (%)
1	*Koelreuteria paniculata*	18.7	28–30	0.91
2	*Rhus typhina*	13.3	22	1.0
3	*Acacia farnesiana*	8.3	23	0.4
4	*Albizzia julibrissin*	9.9	19–24	0.9

Table 1.
The oil content (%) of 4 plants, using mechanical oil extraction and Soxhlet extraction methods.

Peak no.	Wave number (cm^{-1})	Group attribution	Vibration type	Absorption Intensity
1	3465	-OH	Stretching	Weak
2	3006	=C-H	Stretching	Strong
3	2925	$-CH_2$	Asymmetric stretching vibration	Strong
4	2854	$-CH_2$	Symmetric stretching vibration Stretching	Strong
5	1743	-C=O	Shear type vibration	Strong
6	1641	$-CH_2$	Bending vibration	Middling
7	1361	$-CH_3$	Symmetric stretching vibration,	Middling
8	1170	C-O-C	Anti-stretching vibrations	Middling
9	1016	C-O-C	Vibration	Weak
10	723	$-CH_2$	Plane rocking vibration	Weak

Table 2.
FT-IR data presenting various functional groups in FAMEs.

Integration value	Chemical Shift ppm	Multiplicity	Inferences
3	0.89	Multiplet	CH_3 is attached to aliphatic group.
16	1.30	Multiplet	Long aliphatic chain is present.
2	1.62	Quartet	CH_2 group is attached with terminal CH_3.
3	2.04	Multiplet	CH_2 of long chain aliphatic (Saturated) group.
2	2.30	Triplet	CH_2 group is attached with CH of long aliphatic (Unsaturated/ olefinic group).
1	2.77	Triplet	CH group is attached with electron withdrawing carbonyl group.
3	3.66	Singlet	Methoxy(OCH_3)group attached with electron withdrawing carbonyl group.
3	5.34	Multiplet	Olefinic hydrogen of long chain unsaturated aliphatic group

Table 3.
^1H NMR spectroscopic data depicting chemical composition of various methyl esters in biodiesel (FAMES) samples.

Peak No:	Peak area/ region/ ppm	Identified compound	Chemical structure
1	14.07	Terminal methyl carbon	$-CH_3$
2	22.55–34.09	Methylene carbon	$-CH_2$
3	51.37	Methoxy carbon	$-OCH_3$
4	127.91–130.19	Olefinic carbon	C=C
5	174.24	Carboxyl carbon of ester	$-COOCH_3$

Table 4.
^{13}C NMR spectroscopic data depicting the chemical shift values matching to various structural features in FAMEs.

3.4 GC–MS procedure

The outcome of biodiesel in our studies was evaluated by GCMS (QP2010SE, Shimadzu, Japan), furnished with a capillary column: PEG-20 M (30 m × 0.32 mm × 1 μm film thickness). Helium gas flow rate 1.2 mL/min; split ratio 40:1; the injector temperature and injection volume were 220°C and 1 uL; Furnace heat up mode was 100°C for 1 min, then from 100°C rises to 210°C at the increase rate of 10°C/min. Sensor heat mode was 210°C, and then for 20 min, the temperature was continuing at 210°C; ion source temperature of 200°C; for electron impact 70 eV ionization mode used; mass range of 35–500 m/z. The FAMEs of all plant sources were identified with the mass spectrometry fragmentation design provided by the GCMS system software, as matched with those stored in the mass spectrometry library NIST14, and their fatty acid identity was further verified by matching with known standards and values [39–42].

The comparative GC based identified FAMEs major compositions (%) of prepared biodiesel from four non-edible plant sources is given in **Table 5**.

3.5 ICP-OES procedure for elemental analysis in biodiesel

Inductively Coupled Plasma Spectrometer (Spectro-blue, Germany) and Elemental Analyzer (Vario EL CUBE, Germany) were used for the presence of metals in the biodiesel. For the ICP-OES test, 1 g of oil sample was taken for incinerating. The ashing process involved an increase in the oven temperature to 200°C in one hour; then the heat levels were mainatained upto 500 °C for 2 h, and finally

FAMEs major compositions (%)	Plant species			
	Koelreuteria paniculata	*Rhus typhina*	*Acacia farnesiana*	*Albizzia julibrissin*
C16:0	9.7	14.0	6.85	10.598
C16:1	—	—	—	—
C18:0	1.8	3.2	2.36	2.12
C18:2	25.5	47.2	12.13	12.030
C18:3	3.6	1.1	1.23	—
C20:0	2.4	0.8	—	—
C20:1	48.5	0.5	—	—
C21:1	—	—	—	—

Table 5.
GC based identified FAMEs major compositions (%) of prepared biodiesel.

to 800°C for 5 h. The ash was dissolved in 10 mL of 2% HNO_3. The prepared sample was used for elements finding and concentration test of the biodiesel.

The account of ICP-OES comparative element concentrations of 4 non edible oil plant species is given in **Table 6**.

3.6 Elemental analyzer (EA) procedure for elemental analysis

The element analyzer (Vario EL CUBE, Germany) was used to detect the H, N, C and O concentrations of biodiesel obtained from plant sources [39–42]. About 0.5 mL of biodiesel, 3 mL of concentrated HCl and 1 mL of nitric acid were taken in a tube and kept them at rest for 10–15min, to dissolve the oil in the solution. Fresh reagents can be used for sample preparation. The aqua regia amount was twice than the sample. About 1 mL of prepared solution was taken in a new tube and added deionized water making it up to 5 mL. The technique was repeated for 2–3 times until the sample appeared as clear and vivid and ready for evaluation of C, H, N, and O concentrations.

The comparative account of elemental analysis of biodiesel obtained from 4 non edible oil plant species is given in **Table 7**.

3.7 Physiochemical properties of biodiesel seed oil from four non edible oil plant species

The comparative account of physiochemical properties of biodiesel seed oil obtained from four non edible oil plant species is given in **Table 8**.

Elements	Concentration (μg/g)				
	Petro-diesel	KPOB	RTOB	AFOB	AJOB
Sb	—	5744.02	7845.2	—	—
Na	868.3	5456.2	70.29	868.3	868.3
K	213.3	6.14		213.3	213.3
Cr	2.5	1246.8	3.41	—	—
Ni	12.4	658.36	46.98	12.4	12.4
Al	—	346.87	55.56	—	—
Sn	—	378.61	838.6	—	—
Mn	1.5	92.05	5.58	1.5	1.5
Ti	—	64.40	−124.11	—	—
Li	1.6	43.93	105.5	1.6	—
V	—	42.79	15.42	—	—
Cu	99.6	24.63	31.82	—	—
Ca	21.4	14.90		21.4	21.4
Mg	35.6	32.10	32.74	35.6	35.6
Bi	—	19.90	29.05	—	—
Zn	9.5	13.08	49.89	9.5	—
Co	21.2	10.69	7.96	21.2	21.2
Cd	—	—	1.59	—	—

Abbreviations: Koelreuteria paniculata Biodiesel Oil (KPOB); Rhus typhina Biodiesel Oil (RTOB); Acacia farnesiana Biodiesel Oil (AFOB); Albizzia julibrissin Biodiesel Oil (AJOB).

Table 6.
ICP-OES element concentrations of 4 non edible oil plant species.

Ultimate analysis	KPBD	RTBD	AFBD	AJBD
C%	72.54	74.89	76.37	76.86
H%	12.73	413.02	13.34	11.25
N%	2.73	1.97	2.18	2.03
O%	12		8.11	9.86
HHV	23.39		23.39	23.39

Abbreviations: Koelreuteria paniculata Biodiesel Oil (KPOB); Rhus typhina Biodiesel Oil (RTOB); Acacia farnesiana Biodiesel Oil (AFOB); Albizzia julibrissin Biodiesel Oil (AJOB).

Table 7.
Elemental analysis of biodiesel from 4 non edible oil plant species.

Parameters	EN 14214	ASTM D-6751	Petro-diesel	KPOB	RTOB	AFOB	AJOB
Oil contents (wt. %)	—	—	—	28–30	20–22	23	19–24
Density @ 15°C (g/cm^3)	0.86–0.90	0.86–0.90	0.809	0.879	0.879	0.831	0.842
Kinematic viscosity @ 40°C (mm^2/s)	3.5–5.0	1.9–6.0	1.3–4.1	6.21	6.3	5.32	3.75
Flashpoint (°C)	Min. 120	Min. 130	60–80	147	168	158	160
Ignition value	—	—	—	175	—	—	
Acid value ($mg\ KOH/g^{-1}$)	Max. 0.50	Max. 0.5	—	0.07		0.40	
Saponification value ($mg\ KOH/g^{-1}$)	—	—	—	176.4	175.6	174.8	180.4
Iodine value ($g\ I_2/100\ mg$)	Max. 120	Max. 120	—	80.7	85	142.5	118.5
Refractive index @ 20 °C	—	—	—	1.4901	—	—	
Cloud point (°C)	—	—	−15–5	2	7	7	9
Pour point (°C)	—	—	−2.0	−30	−11	−28	−12
Fire point (°C)	—	—	—		198	189	190
Cetane number	Min. 51	Min. 47	49.7	51	—	52	58
Free fatty acid (%)	—	—	—	0.91	1.0	0.4	0.9
HHV (MJ/kg)	—	—	—	23.39	23.73	—	—
Ash content (g/100 g)	—	—	—	0.002	0.3	0.002	0.003
Specific gravity @15 (°C)	—	—	—	0.88	0.855	0.831	0.842
Cold filter plug point (°C)	Max.19	Max.19	−16	−18	14	−25	3
Sulphated ash content (wt.%)	Max.0.02	—	—	0.003	—	—	
Oxidation stability (110°C, h)	Min. 6	Min. 3	25.8	—	18.3	−1.86	4.71

Abbreviations: Koelreuteria paniculata Biodiesel Oil (KPOB); Rhus typhina Biodiesel Oil (RTOB); Acacia farnesiana Biodiesel Oil (AFOB); Albizzia julibrissin Biodiesel Oil (AJOB).

Table 8.
Physiochemical properties of biodiesel (FAMEs) samples.

4. Discussion

The great potential exists for using non-edible seed oil as biodiesel, which has been highlighted in this chapter. The plant sources explored were *Koelreuteria paniculata*, *Rhus typhina*, *Acacia farnesiana* and *Albizzia julibrissin*.

Biodiesel can be synthesized from vegetable oils, animal fats or algae oils [51]. The Soxhlet extraction allows the sample to repeatedly bring into contact with fresh portions of extracting materials, hence facilitating the equilibrium. It remains at relatively high temperature and no filtration is required [52]. In the present study, *Koelreuteria paniculata* produced highest biodiesel oil content by Soxhlet extraction (28–30%) followed by the *Albizzia julibrissin* (19–24%), *Acacia farnesiana* (23%), *Rhus typhina* (20–22%) than the mechanical extraction.

The biodiesel production largely depends upon the choice of appropriate, cost-effective and environment friendly catalysts [53] based on the nature of oil which greatly help in transesterification of oil. Density optimization is a vital biodiesel factor having huge impact on fuel quality as well the cost incurred [54]. The density for all plant sources ranged from 0.83–0.87 @ 15°C (g/cm3). The kinematic viscosity ranged from 3.75–6.3 (mm^2/s) among all the plant sources. Our results show that densities of all plant sources were within the ASTM (D6751) and (EN14214) standards.

The 1H-NMR procedure is used to evaluate the un-saturation and "residual" fatty acid composition [55]. In our study, the biodiesel 1H NMR (300 MHz) spectrum was noted with a cycle delay of 1.0 s, and eight times scans with a pulse duration of 30°. 13C NMR spectroscopic data represented the chemical shift values matching to various structural features in FAMEs.

The metals such as Cu, Co, Fe, Mn and Ni are known to catalyze oxidative degradation reactions in vegetable oils and biodiesel [56]. The presence of some metals, such as Al, Ca, Cr, Cu, Fe, Mg, Pb, V and Zn, among other, in fuel is undesirable due to their release into the atmosphere upon fuel combustion [57]. *Koelreuteria paniculata* had highest Na (5456.2), Cr (1246.8), Ni (658.36), and Al (346.87) elemental concentrations (μg/g) than other plant sources. Moreover, Al, Cr, Sn, V, Cu, Bi, Cd were not detected in *Acacia farnesiana* and *Albizzia julibrissin* and both showed comparable Zn, K, Ni, Mn, Ca, and Co concentrations with petro diesel. The metal elements in biodiesel result into engine degradation, operational as well as cause environmental pollution [58]. In our study, the elemental percent of C, H, N, and O of biodiesel ranged from 72.54–76.86, 11.25–13.34, 1.97–2.73, and 9.86–12, respectively.

Flash point of fuels is imperative to determine the prerequisites for transportation and storage temperature [59]. The flash point was highest in the biodiesel oil of *Rhus typhina* (168°C),*followed by Albizzia julibrissin* (160°C), *Acacia farnesiana* (158°C), *Koelreuteria paniculata* (147°C) which was comparable to EN 14214 (Min. 120°C), ASTM D-6751 (Min. 130°C) petro diesel (60–80°C) standards.

The main shortcomings of biodiesel are related to the low-temperature performance and oxidation stability. The oxidation stability (110°C, h) was recorded as 18.3 *in Rhus typhina*, 4.71 in *Albizzia julibrissin*, –1.86 in *Acacia farnesiana* and these fall in the recommended values of EN 14214 (Min. 6), ASTM D-6751 (Min. 3) and petro diesel (25.8). This highlights that the saturated fraction of biodiesel has a positive effect on its stability as in the case of our results, but the long-chain saturates may compromise fuel performance at lower temperatures [60].

Metal-containing fuel additives and un-removed catalysts are the major contributors of sulfated ash [61]. The sulphated ash content (%) was not detected in *Rhus typhina*, *Acacia farnesiana*, *Albizzia julibrissin* and it was 0.003 in *Koelreuteria paniculata* as compared to the EN 14214 (Max.0.02) standards, hence these plant sources can be used for biodiesel production.

5. Conclusions

The physico-chemical characterizations of the biodiesel, i.e., flash point, pour point, cloud point, and density, have been found within the ASTM (D6751) and (EN14214) standards. The values of fuel properties were comparable with mineral diesel. FT-IR, NMR, and GC–MS analysis established the total conversion of crude oil to FAMEs. The elemental analysis of biodiesel ensured the feasibility for environment friendly usage. These sources have been optimized through optimization of transesterification reactions: oil to methanol ratio (6:1), a potassium hydroxide concentration (3.0%), temperature (65°C), stirring rate (700 rpm) and reaction time (60–80 min). The highest biodiesel yield was obtained from *Albizzia julibrissin* (98%) followed by *Acacia farnesiana* (96%), *Koelreuteria paniculata* (95.2%) and *Koelreuteria paniculata* (93.33%). FAMEs of all sources were compatible to ASTM (D6751) and (EN14214) standards. These non-edible plant seeds offer a cheap source of renewable energy. These plants can be easily grown on barren and wastelands and contribute to efficient biodiesel production to curtail the energy crisis. Keeping these findings in preview, we can assert that biodiesel obtained from these non-edible seeds has a huge potential as an alternative to petroleum diesel and can be efficient renewable source of fuel.

Conflict of interest

The authors declare that they have no known competing financial interests or personal relationships that could have appeared to influence the work reported in this chapter.

Authors' contributions

Inam Ullah Khan prepared the first draft of the chapter based on his published research work and other relevant literature. Syed Aftab Hussain Shah edited, formatted the entire chapter and worked on bibliography.

Ethical statement

This is to certify that chapter has adhered to the research ethics.

Author details

Inam Ullah Khan[1,2]* and Syed Aftab Hussain Shah[3]

1 Key Laboratory of Advanced Energy Materials Chemistry (Ministry of Education), Renewable Energy Conversion and Storage Center, College of Chemistry, Nankai University, Tianjin, P.R. China

2 Department of Chemical and Biological Engineering, Arts Tower Western Bank, The University of Sheffield, Sheffield, Manchester, UK

3 Pakistan Scientific and Technological Information Centre, Quaid-i-Azam University Campus, Islamabad, Pakistan

*Address all correspondence to: khanqau123@yahoo.com

References

[1] Röllin H. Evidence for health effects of early life exposure to indoor air pollutants: What we know and what can be done. Clean Air J. 2017;27(1).

[2] Xue B-j, Luo J, Zhang F, Fang Z. Biodiesel production from soybean and Jatropha oils by magnetic CaFe2O4–Ca2Fe2O5-based catalyst. Energy. 2014;68:584-91.

[3] Samadi S, Karimi K, Behnam S. Simultaneous biosorption and bioethanol production from lead-contaminated media by Mucor indicus. Biofuel Res. J. 2017;4(1):545.

[4] Knothe G, Krahl J, Van Gerpen J. The Biodiesel Handbook: Elsevier; 2015.

[5] Garcez CAG, de Souza Vianna JN. Brazilian biodiesel policy: social and environmental considerations of sustainability. Energy. 2009;34(5):645-654.

[6] Živković SB, Veljković MV, Banković-Ilić IB, Krstić IM, Konstantinović SS, Ilić SB, et al. Technological, technical, economic, environmental, social, human health risk, toxicological and policy considerations of biodiesel production and use. Renew. Sustain. Energy Rev. 2017;79:222-247.

[7] Tremblay AY, Montpetit A. The in-process removal of sterol glycosides by ultrafiltration in biodiesel production. Biofuel Res. J. 2017;4(1):559-564.

[8] Sharma Y, Singh B, Upadhyay S. Advancements in development and characterization of biodiesel: a review. Fuel. 2008;87(12):2355-2373.

[9] Song H, Quinton KS, Peng Z, Zhao H, Ladommatos N. Effects of oxygen content of fuels on combustion and emissions of diesel engines. Energies. 2016;9(1):28.

[10] Nakano M, Okawa K. Study of oxygen-containing hydrocarbons in exhaust emission from a spark ignition combustion engine. Int. J. Engine Res. 2014;15(5):572-580.

[11] Mwangi JK, Lee W-J, Chang Y-C, Chen C-Y, Wang L-C. An overview: Energy saving and pollution reduction by using green fuel blends in diesel engines. Appl. Energy. 2015;159:214-236.

[12] Murayama T. Evaluating vegetable oils as a diesel fuel. 1994;5:1138-1145.

[13] Knothe G, Gerpen J, Krahl J. The Biodiesel Handbook. Champaign-Illinois: AOCS Press; 2005. p. 89-90.

[14] Verma D, Raj J, Pal A, Jain M. A critical review on production of biodiesel from various feedstocks. J. Scient. Innov. Res. 2016;5(2):51-58.

[15] Roeges NP. A guide to the complete interpretation of infrared spectra of organic structures: J. Wiley; 1994. p. 1-356.

[16] Akinsiku AA. Biodiesel fuel from differently sourced local seed oils: characterization, effects of catalysts, total glycerol content and flow rates. Int. J. Scient. Eng. Res. 2013;4(6):654-660.

[17] Otera J. Transesterification. Chemical Rev. 1993;93(4):1449-1470.

[18] Demirbas A. Biodiesel production from vegetable oils via catalytic and non-catalytic supercritical methanol transesterification methods. Prog. Energy Combust. Sci. 2005;31(5-6):466-487.

[19] Ma F, Hanna MA. Biodiesel production: a review. Bioresour. Technol. 1999;70(1):1-15.

[20] Körbitz W. Biodiesel production in Europe and North America, an

encouraging prospect. Renew. Energy. 1999;16(1-4):1078-1083.

[21] Sharma Y, Singh B. Development of biodiesel: current scenario. Renew. Sustain. Energy Rev. 2009;13(6-7):1646-1651.

[22] Wazilewski WT, Bariccatti RA, Martins GI, Secco D, de Souza SNM, Rosa HA, et al. Study of the methyl crambe (*Crambe abyssinica* Hochst) and soybean biodiesel oxidative stability. Ind. Crops Prod. 2013;43:207-212.

[23] Gui MM, Lee K, Bhatia S. Feasibility of edible oil vs. non-edible oil vs. waste edible oil as biodiesel feedstock. Energy. 2008;33(11):1646-1653.

[24] Vargas-Lopez J, Wiesenborn D, Tostenson K, Cihacek L. Processing of crambe for oil and isolation of erucic acid. J. Am. Oil Chem. Soc. 1999;76(7):801-809.

[25] McKendry P. Energy production from biomass (part 2): conversion technologies. Bioresour. Technol. 2002;83(1):47-54.

[26] Aliyu B, Agnew B, Douglas S. Croton megalocarpus (Musine) seeds as a potential source of bio-diesel. Biomass and Bioenergy. 2010;34(10):1495-1499.

[27] Atapour M, Kariminia H-R. Characterization and transesterification of Iranian bitter almond oil for biodiesel production. Appl. Energy. 2011;88(7):2377-2381.

[28] Wang L, Yu H. Biodiesel from Siberian apricot (*Prunus sibirica* L.) seed kernel oil. Bioresour. Technol. 2012;112:355-8.

[29] Nehdi IA, Sbihi HM, Al-Resayes SI. *Rhazya stricta* Decne seed oil as an alternative, non-conventional feedstock for biodiesel production. Energy Convers. Manag. 2014;81:400-406.

[30] 30.Reshad AS, Tiwari P, Goud VV. Extraction of oil from rubber seeds for biodiesel application: Optimization of parameters. Fuel. 2015;150:636-644.

[31] Fadhil AB, Aziz AM, Al-Tamer MH. Biodiesel production from *Silybum marianum* L. seed oil with high FFA content using sulfonated carbon catalyst for esterification and base catalyst for transesterification. Energy Convers. Manag. 2016;108:255-265.

[32] Al-dobouni IA, Fadhil AB, Saeed IK. Optimized alkali-catalyzed transesterification of wild mustard (*Brassica juncea* L.) seed oil. Energy Sources, Part A: recovery, utilization, and environmental effects. 2016;38(15):2319-2325.

[33] Rashed M, Kalam M, Masjuki H, Mofijur M, Rasul M, Zulkifli N. Performance and emission characteristics of a diesel engine fueled with palm, jatropha, and moringa oil methyl ester. Ind. Crops Prod. 2016;79:70-76.

[34] Sinha P, Islam MA, Negi MS, Tripathi SB. Changes in oil content and fatty acid composition in *Jatropha curcas* during seed development. Ind. Crops Prod. 2015;77:508-510.

[35] Foidl N, Foidl G, Sanchez M, Mittelbach M, Hackel S. *Jatropha curcas* L. as a source for the production of biofuel in Nicaragua. Bioresour. Technol. 1996;58(1):77-82.

[36] Bhatti HN, Hanif MA, Qasim M. Biodiesel production from waste tallow. Fuel. 2008;87(13-14):2961-2966.

[37] Chisti Y. Biodiesel from microalgae. Biotechnology advances. 2007;25(3):294-306.

[38] Chen Y-H, Chen J-H, Luo Y-M. Complementary biodiesel combination from tung and medium-chain fatty acid oils. Renew. Energy. 2012;44:305-310.

[39] Khan IU, Yan Z, Chen J. Production and characterization of biodiesel derived from a novel source *Koelreuteria paniculata* seed oil. Energies. 2020;13(4):791.

[40] Khan IU, Yan Z, Chen J. Optimization, transesterification and analytical study of *Rhus typhina* non-edible seed oil as biodiesel production. Energies. 2019;12(22):4290.

[41] Khan IU, Yan Z, Chen J, Chen H, Shah SAH. *Acacia farnesiana* seed oil: a promising source of biodiesel production. Int. J. Scient. Eng. Res. 2021;12(1):277-295.

[42] Khan IU, Yan Z, Chen J, Chen H, Shah SAH. *Albizzia julibrissin*: a non-edible seed oil source for biodiesel production. Int. J. Scient. Eng. Res. 2021;12(1):327-341.

[43] Martins-Noguerol R, DeAndrés-Gil C, Garcés R, Salas JJ, Martínez-Force E, Moreno-Pérez AJ. Characterization of the acyl-ACP thioesterases from *Koelreuteria paniculata* reveals a new type of FatB thioesterase. Heliyon. 2020;6(10):e05237.

[44] Ruan C-J, Xing W-H, da Silva JAT. Potential of five plants growing on unproductive agricultural lands as biodiesel resources. Renew. Energy. 2012;41:191-199.

[45] Zhang F, Ren H, Tong G. Production of biodiesel from the extract of the sumac fruit cluster. Cellul. Chem. Technol. 2018;52:1275-1279.

[46] Clarke HD, Seigler DS, Ebinger JE. *Acacia farnesiana* (Fabaceae: Mimosoideae) and related species from Mexico, the southwestern US, and the Caribbean. Syst. Bot. 1989:549-564.

[47] Weber E, editor. Invasive plant species of the world: a reference guide to environmental weeds: Cabi; 2017. DOI 10.1079/9781780643861.0000

[48] Vines RA. Trees, shrubs and woody vines of the southwest. 5th ed.University of Texas at Austin; 1960. 1104 p.

[49] Williams R, Hanks S. Hardwood nursery guide. Slightly rev., July 1994. Agriculture handbook (USA). 1976. p. 473.

[50] Cheatham S, Johnston M, Marshall L. useful wild plants of Texas, the southeastern and southwestern United States, the southern plains and northern Mexico, vol. I. HerbalGram. 1998.

[51] Ozcanli M, Gungor C, Aydin K. Biodiesel fuel specifications: A review. Energy Sources, Part A: recovery, utilization, and environmental effects. 2013;35(7):635-647.

[52] De Castro ML, Priego-Capote F. Soxhlet extraction: Past and present panacea. J. Chromatogr. A. 2010;1217(16):2383-2389.

[53] Gupta J, Agarwal M, Dalai A. An overview on the recent advancements of sustainable heterogeneous catalysts and prominent continuous reactor for biodiesel production. J. Ind. Eng. Chem. 2020;88:58-77.

[54] Pratas MJ, Freitas SV, Oliveira MB, Monteiro SC, Lima ÁS, Coutinho JA. Biodiesel density: experimental measurements and prediction models. Energy Fuels. 2011;25(5):2333-2340.

[55] Knothe G, Kenar JA. Determination of the fatty acid profile by 1H-NMR spectroscopy. Eur. J. Lipid Sci. Technol. 2004;106(2):88-96.

[56] Choe E, Min DB. Mechanisms and factors for edible oil oxidation. Compr. Rev. Food Sci. F. 2006;5(4):169-186.

[57] Wang Y-F, Huang K-L, Li C-T, Mi H-H, Luo J-H, Tsai P-J. Emissions of fuel metals content from a diesel vehicle engine. Atmos. Environ. 2003;37(33):4637-4643.

[58] Sorate KA, Bhale PV. Biodiesel properties and automotive system compatibility issues. Renew. Sustain. Energy Rev. 2015;41:777-798.

[59] Santos SM, Nascimento DC, Costa MC, Neto AM, Fregolente LV. Flash point prediction: Reviewing empirical models for hydrocarbons, petroleum fraction, biodiesel, and blends. Fuel. 2020;263:116375.

[60] Wang W, Li F, Li Y. Effect of biodiesel ester structure optimization on low temperature performance and oxidation stability. J. Mater. Res. Technol. 2020;9(3):2727-2736.

[61] American Society for Testing Materials. ASTM 6751-Standard specification for biodiesel fuel (B100). Blend stock for distillate fuels. ASTM, West Conshohocken PA, USA; 2002.

Chapter 4

Plant-based Vaccines: The Future of Preventive Healthcare?

Sinan Meriç, Tamer Gümüş and Alp Ayan

Abstract

Infectious diseases threatened humankind countless times through history, when knowledge on microorganisms was absent and medical capabilities were limited. Pandemics and outbreaks caused death of millions, brought empires to their knees and even wiped some ancient civilizations. In "modern" days, despite of improved medical application, sanitary precautions and effective medicines, infectious diseases are still cause of more than 54% of total mortality in developing countries. Millions of people are protected from the infectious diseases annually as a result of mass immunization campaigns. Nevertheless, novel diseases as COVID-19, MERS-CoV, avian influenza, Ebola, Zika and possible future infections require dynamic vaccine research and investment. Along with all the advantages of vaccines, there are several limitations regarding cost, biosafety/biosecurity, storage, distribution, degradation topics. Plant-based vaccine production for humans and animals has been under serious consideration to overcome some of these limitations. Nowadays, plant biotechnology brought new insight to vaccines research through gene transfer strategies to plants and improvements in amount, isolation and purification and addition of adjuvant for production of recombinant vaccine antigens in plants. Recombinant vaccines can undeniably offer us new standards and legal regulations to be introduced for the development, approval, authorization, licensing, distribution and marketing of such vaccines. The aim of this chapter is to exploit uses, methods and advantages of recombinant DNA technology and novel plant biotechnology applications for plant-based vaccine research in respect to existing infectious diseases.

Keywords: Plant-based vaccine, recombinant protein, virus-like particles, transgenic plant, molecular farming, oral vaccine, chloroplast transformation, transient expression, stable nuclear transformation, COVID-19

1. Introduction

There are many breakthrough events during ongoing human civilization process. All of these individual events contributed the process in different significance. Nevertheless, agricultural development and animal domestication significantly accelerated all the other developments due to the fact that they saved people from daily boundary of finding nutrition and allowed more spare time for socialization and thinking.

Advantages of settled life style increased population in early cities rapidly. Ancient cities as Rome, Athens, Fayum varied in population between a hundred thousand to a million in various eras [1, 2]. Lacking the knowledge of microorganisms, hygiene and sanitation precautions, underdeveloped sewer systems and living so close to domesticated animals and people resulted to the rise of "civilization pathogens". It is hypothesized that virulent pathogens were present but not persistent due to the limited population in human communities before agricultural development and urbanization. Most of the animals tend to live together in herds. Even though the herd lifestyle is very suitable for the transmission of pathogens, there were limited contact between humans and animals since hunting was the only viable way. Developments in agriculture and animal domestication lifted that barrier and allowed animal diseases to be transmitted to humans more frequently in higher population densities. Major fatal human diseases as measles, tuberculosis, smallpox, influenza, pertussis, and falciparum malaria are linked to early domesticated animals via phylogenetic analysis [3]. Throughout the history of human civilization there are several outbreaks of pandemic diseases which shaped the world socially, economically and politically (**Table 1**).

In present day, vaccines are the vital part of the preventive healthcare globally. Many of the once deadly diseases are not present for decades, since the mass vaccination campaigns were applied worldwide. Based on their production method and protection mechanisms, vaccines are categorized under several classes including

Common Name	Year(s)	Cause	Estimated Death (Million)
Plague of Athens	430 B.C.	*Salmonella enterica*	0.1
Antonine Plague	165–180	Small pox (*Variola major* and *Variola minor*) Measles (*Measles morbillivirus*)	5
Plague of Justanian	541–542	*Yersinia pestis*	30–50
Black Death	1347–1352	*Yersinia pestis*	75–200
New World Smallpox	1520 - NA	*Variola major* and *Variola minor*	25–55
Italian Plague	1629–1631	*Yersinia pestis*	1
Great Plague of London	1665	*Yersinia pestis*	0.075–0.1
Third Plague	1885	*Yersinia pestis*	12
Russian Flu	1889–1890	A/H3N8, A/H2N2, or coronavirus OC43 (uncertain)	1
Yellow Fewer	Late 1800s	RNA virus from *Flavivirus* genus	0.15
Spanish Flu	1918–1920	Influenza strains of A/H1N1	50
Asian Flu	1957–1958	Influenza A virus subtype H2N2	1
Hong Kong Flu	1968–1970	H3N2 strain of the influenza A virus	1
HIV-AIDS	1981-ongoing	*Lentivirus*	35 and counting
SARS	2002–2003	Coronavirus (SARS-CoV-1)	<0.001
Swine Flu	2009	H1N1 influenza virus	0.2
MERS	2015	Coronavirus (MERS-CoV)	<0.001
Ebola	2014–2016	Ebolaviruses	0.011
Covid-19	2019-ongoing	Coronavirus (SARS-CoV-2)	2.52 and counting

Table 1.
Major outbreak throughout the history of human civilizations [4].

live attenuated, inactive (killed whole organism), toxoid, subunit (purified native or recombinant protein, polysaccharide or peptide), virus-like particle, outer membrane particle, protein-polysaccharide conjugate, viral vectored, nucleic acid, bacterial vectored, antigen-presenting cell vaccines [5, 6]. Despite the various new approaches to the vaccine production, most of the vaccines which are applied in immunization programs are either live attenuated, inactive or subunit vaccines. WHO (World Health Organization) vaccine-preventable diseases: monitoring system [7] offers important and comparable data on this topic based on all countries. Even so, there are public concerns over the topics as age and schedule of vaccination, common (injection site pain, redness and swelling, fever, malaise, headache) and rare side effects (anaphylaxis, idiopathic thrombocytopenic purpura, narcolepsy, autism), immunodeficiency or antigenic overload issues. As in all daily life matters, misinformation on these topics and issues in social media and search engines greatly challenges vaccine production methods and public acceptance, although the scientific evidences prove the contrary.

As illustrated in **Figure 1**, one of the greatest challenges in vaccine research is based on logistics and distribution. Along with post-production purification and packaging issues, logistics of traditional vaccines under cold chain conditions in limiting shelf-life durations exhibit certain difficulties [6]. Especially in underdeveloped countries, lack of health infrastructure and required conditions threaten overall process. Moreover, world once again faced the same dilemma in Covid-19 pandemic as 16 of total 256 vaccine candidates passed to phase III trials by February, 2021 [8]. These manufacturers announced their plans to produce 10 billion doses (at least 2 doses are required for immunity in most) until the end of 2021 in their best estimations. Apparent insufficient annual production capacity leads priority groups to be formed for the early products. As in this latest ongoing example, inequity in access to vaccines is always a major issue [9].

Leading Covid-19 vaccine candidates are mainly developed by private/industry association by 72%. Remaining 28% is consisted of academic, public and non-profit organizations. Also, there are bigger multinational vast vaccine manufacturer companies as Janssen, Sanofi, Pfizer and GlaxoSmithKline which may ensure tolerating lack of large scale vaccine manufacturing inexperience and capacities of these relatively smaller organizations [10]. Even though, commercial viability

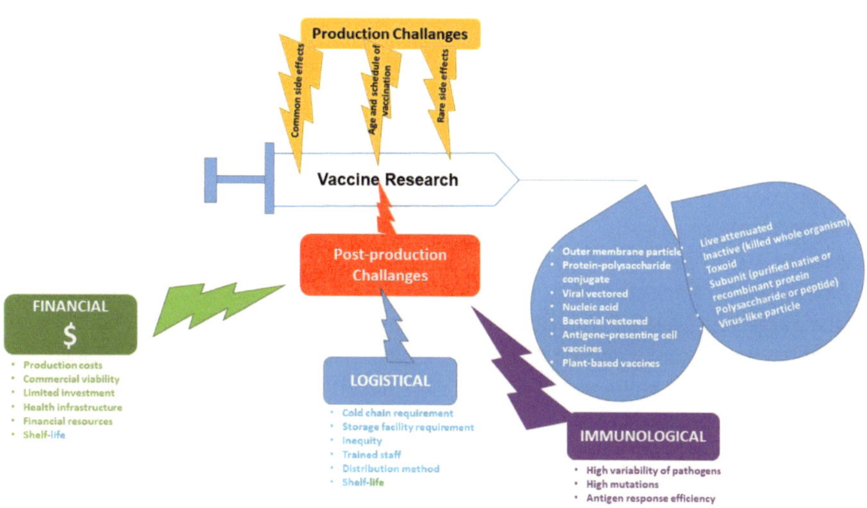

Figure 1.
Main global challenges in vaccine research.

apparently is not an issue for potential Covid-19 vaccine, it is a real drawback for many diseases. These diseases are mostly have devastating effects on restricted local areas as poor communities. In case of such rare infection diseases, development costs offset potential income. Vaccines against this kind of diseases as Ebola, which multinational manufacturers hesitate to invest due to commercial viability, are called 'orphan vaccines' [11]. More profitable vaccine production methods may withdraw hesitation over these diseases which are only producible with government assistance and still have high mortality rates regionally.

Another and probably the most challenging factor in vaccine research is based on immunological issues. Commercially viable vaccine targets for diseases like HIV, gonorrhea (*Neisseria gonorrhoeae*), syphilis (*Treponema pallidum*), malaria or seasonal influenza are generally caused by highly variable pathogens. These pathogens present variation both in and between host variations which emerge difficulty to identify common antigen target for immunization by vaccine. Some people produce natural antibodies against more conserved antigens of pathogens and have enhanced immunity but targeting these conserved antigens by vaccine induction has not been achieved for many of these diseases yet. Protection efficacy is a major issue due to the immunological variations even in post-production of the vaccines [6]. Even so, global reports over novel mutations of the virus will require re-evaluation of efficacy values and raise suspicion in public for existing vaccines. In some case as RTSS vaccine which is licensed for malaria disease, efficacy is lower as 30–40%. Therefore, considering all immunological factors together suggests long development time of traditional vaccines fail to satisfy rapid, flexible and upscale production requirements [6].

Following the rapid development of biotechnology and bioinformatics, precise genomic and proteomic target identification methods and instruments are emerged. Therefore, knowledge on structural biology and immunology is mostly available for many infectious diseases. Deciding vaccine production methods is based on delivery, immunogenicity, production capacity and speed, transport requirements and shelf life and economic viability. Along with the methods as viral vectored vaccines, nucleic acid-based vaccines, RNA vaccines, outer membrane vesicles, plant based vaccines present promising contribution to the field. Experimental and commercial applications of plant based vaccines will be evaluated further in this chapter in perspective of future preventive healthcare alternative.

2. Plant-based vaccine production

Especially in the last two decades, the expression studies of vaccine antigens in plants have been accelerated with the developments in the production of recombinant proteins in plants and it has been provided possibility to design effective plant-based vaccines against many diseases. In this process, both developments in transgenic approaches and transformation methods and improvements in various areas such as promoter selection, codon optimization, plastid transformation for increasing yield of recombinant protein have paved the way for the production of vaccine antigens. The significant increase in the world population and the emergence of epidemic and pandemic diseases cause demands that exceed the vaccine production capacity. However, the success of national vaccine programs is marred by both high cost-per-dose of producing vaccines and the limitations in the distribution of vaccine. In addition, there is a risk of exposure to dangerous pathogens caused by injection procedures during vaccination, resulting in diseases like HIV, hepatite C that can be transmitted through blood. Moreover, risk of contamination of other viruses such as SV40 and foamy virus, which can cause disease in humans and animals, is another important factor that should be evaluated in terms of health, although it depends on the nature of the vaccine (attenuated or inactivated), the titer of the contaminant, the degree of inactivation and pathogenicity [12]. When all these disadvantages are examined, it is seen that the use of plant systems for vaccine production has the potential to provide a biotechnological solution, considering that it can provide high-scale production and reduce the cost-per-dose and minimize the problems that may occur during vaccine production and distribution. While plants produce complex proteins similar to other eukaryotic systems, they can fold and modify these proteins post-translationally. However, they contain minor differences in glucosylation patterns compared to mammalian cells [13].

In general, the technical points taken into consideration for the plant transformation and the production of recombinant protein in plants should also be taken into account in planning for the production of recombinant vaccine antigens in plant systems. Two different systems are used in the production of recombinant proteins in plants known as stable genetic transformation and transient gene expression. Stable nuclear transformation results in stable expression in plant tissues by ensuring the insertion of recombinant DNA into the nuclear genome of the plant cell [13]. In addition, transgenes can stably integrate into the plastid genome other than the nuclear genome. The transfer of recombinant DNA is carried out by using direct and indirect methods and this preference varies according to the target plant species, target genome (nuclear or plastid), gene construct to be introduced. Natural plant pathogens, *Agrobacterium tumefaciens* and *Agrobacterium rhizogenes* are used for indirect gene transfer and are generally preferred for nuclear transformation. Biolistic or microparticle bombardment, which is the most preferred among direct methods, are mostly used in the transformation of the plastid genome and plant species where transformation mediated by *Agrobacterium* species is not applicable. By using plant tissue culture methods with all these transformation methods, a whole transgenic plant can be obtained or plant tissue cultures (callus, hairy root) can be established for recombinant protein production from various plant tissues. In addition, by using plant tissues for recombinant protein production, plant cell cultures also become prominent as an alternative system. Generally, transgenic plants allow large-scale production of recombinant proteins with high expression of introduced gene. This system can also enable the production of multiple recombinant proteins in a single plant by crossing different transgenic plants. On the other hand, development of transgenic plants by stable transformation is relatively more time consuming and also needs improvement due to the low protein expression compared to other plant systems. Another disadvantage is that the transgene may show different profiles in its expression due to positional effect as a result of random entry into the nuclear genome. Moreover, in case of multiple insertions, unstable gene expression and gene silencing may occur. On the other hand, transient expression has many advantages for the expression of genes encoding recombinant proteins in plant tissues. Two processes are particularly prominent for transient transformation in plants named as transient expression of the transgene by *Agrobacterium* infiltration and viral vector-based transient expression. These two processes are based on the expression of transgenes transported by bacteria or virus vectors, and stable integration of the transgene into the genome is not required. The most important advantage of these systems is rapid recombinant protein production. Expression of extra chromosomal transgenes can be detected in 3–4 hours after DNA transfer, while it can reach the maximum expression level in 18–48 hours [14]. Gene expression can be maintained for 10–14 days, afterwards. Transient based expression is generally at a higher level than stable transformation. Plant viral vectors used for viral vector-based transient expression, can be preferred to increase the number of gene copies that can result in a much higher protein yield compared to stable transformation. Various RNA viruses such as tobacco mosaic virus (TMV), cauliflower mosaic virus (CMV), alfalfa mosaic virus (AVM) are used to construct plant viral expression vectors.

2.1 Nuclear transformation

It is ensured that vaccine antigens can be produced in large amounts in the tissues of transgenic plants that are transformed by nuclear transformation, and oral administration of the vaccine becomes possible thanks to the expression in edible plant organs in such as lettuce (*Lactuca sativa*). The gene integrated into nuclear genome can be maintained with transgenic seeds and replanted when needed. Moreover, by crossing different transgenic plants, different transgenes can be included in a single plant, allowing development of different characteristics of the plant in a short time. Transgenic plant lines, depending on the plant species, can be developed in a shorter time by nuclear transformation than chloroplast transformation. Recombinant vaccine antigens produced by nuclear transformation can be targeted to a variety of organelles such as chloroplasts, vacuoles and endoplasmic reticulum owing to signal peptides, and various post-translational modifications can be achieved particularly in the endoplasmic reticulum. On the other hand, the disadvantage of nuclear transformation is that it displays relatively low levels of expression compared to chloroplast transformation and transient expression, as well as position effect and gene silencing [15].

For many years, the nuclear genome has been the main target of plant gene transfer studies, which has enabled the production of recombinant vaccine antigens by nuclear transformation in plants to come to the fore and to be performed relatively easily. Thanks to nuclear transformation in plants, the production of vaccines against many disease factors from enteric bacteria to viruses that threaten human and animal health has been achieved (**Table 2**).

Every day, 200 million people in the world experience health problems due to gastroenteritis. In developing countries, more than 2 million people die annually from such enteric diseases [23]. It has been reported that a multiepitopic protein from the antigens of enterotoxigenic *E. coli*, *S. typhimurium* and *V. parahaemolyticus* bacteria for use against these diseases is expressed in fresh leaf tissue as much as 5.29 $\mu g\ g^{-1}$ in tobacco (*Nicotiana tabacum*) plants by *A. tumefaciens*-mediated transformation. Plant-made LTBentero antigen was found to be immunogenic when administered orally or subcutaneously to BALB/c mice, this antigen was also able to induce specific IgG (systemic) and IgA (mucosal) responses against LTB, ST, and LptD epitopes [23]. Moreover, by regulating the quality processes of transgenic plant-based vaccines, it has been ensured that the quality differences in the production steps are minimized. Especially for this purpose, closed hydroponic plant growing systems were established and oral cholera vaccine (MucoRice-CTB) was produced in accordance with legal regulations within the scope of "Good Manufacturing Practices" without the need for purification [24]. Thus, a plant-based vaccine production system, which can be harvested three times a year and is effective in terms of cost and production, has been implemented. Moreover, Needle- and cold chain free rice-based oral vaccine MucoRice-CTB has been reported to show immunogenicity in humans due to microbiota [25].

In recent years, except advanced monocot and dicot transgenic plants, various moss groups in the plant kingdom have started to be preferred for vaccine production. Env-based HIV (Human Immunodeficiency Virus) multi-epitope protein (poly-HIV) produced in transgenic moss lines has been reported to elicit an immune response in mice as a candidate for subunit vaccine. Moss can be propagated under *in vitro* conditions in accordance with "Good Manufacturing Practices" standards and algae biomass does not have an apparent toxic effect. Therefore, this production system allows immunization with raw moss material. Moreover, mosses

DOI: http://dx.doi.org/10.5772/intechopen.97861

Plant	Pathogen/ Disease	Antigent	Transfer Method	% of Total Soluble Protein or μg of antigen/g fresh biomass	Adjuvant	Immunization	Target Organism	Immun Response	Ref.
Lactuca sativa	Hepatitis B virus	small surface antigen (S-HBsAg)	CaMV 35S promoter/ *Agrobacterium tumefaciens*-mediated	22–63 μg/g FW	With adjuvant (alhydrogel)	Orally-tablet, intramuscularly /total 200 ng	Human	IgA	[16]
Physcomitrella patens	HIV / AIDS	Gp 120 and Gp 41 attached poly-HIV	PEG-Mediated by Protoplast	3.7 μg/g FW	With adjuvant (Freund's complete adjuvant and Freund's incomplete adjuvant)	Subcutaneously/ 34 ng	Human	IgG İnduction	[17]
Nicotiana tabacum	*Mycobacterium tuberculosis*	TBAg-ELP Fusion Protein	CaMV 35S promoter/ *Agrobacterium tumefaciens*-mediated	4% TSB	With adjuvant (Freund's complete adjuvant and Freund's incomplete adjuvant)	Intraperitoneal/ for mice 10 μg, for piglet 360 μg	Human	Induced IgG	[18]
Lactuca sativa	Measles virus	Measles virus hemagglutinin (MV-H)	*Agrobacterium tumefaciens*-mediated	NA	Only intranasal with adjuvant (crude saponins)	Intranasally (40-100 μg-tsb)/ Intraperitonealy/ (400-1000 μg-tsb)/ intramuscularly (50 μg-single dose)	Human (Child)	Induced IgG	[19]

Plant	Pathogen/ Disease	Antigent	Transfer Method	% of Total Soluble Protein or µg of antigen/g fresh biomass	Adjuvant	Immunization	Target Organism	Immun Response	Ref.
Solanum tuberrosum	Infectious bronchitis virus (IBV)	Spike (S) protein	CaMV 35S promoter/ *Agrobacterium tumefaciens*-mediated	2.5 µg/g FTW	Witout adjuvant	Orally 35 g/ intramuscularly 35 g extract	Poultry	Induced chIL-2	[20]
Solanum tuberosum/ Lycopersicon esculentum	Norwalk virus (NV)	Recombinant capsid protein rNV	Dual-enhancer 35S promoter /*Agrobacterium tumefaciens*-mediated	ST:120 µg/g FW LE:150 µg/g FW	Without adjuvant	Orally/ 5 g dry weight	Human	Induced serum IgG and intestinal IgA	[21]
Zea mays	Rabies virus RV	G protein	Maize ubiquitin promoter / transformed by biolistics	25 µg/g FW	Without adjuvant	Orally / 0.5–2 mg x 5 fresh kernels	Human and Dog	NA	[22]
Nicotiana tabacum	*E. coli*, *S. typhimurium*, *V. parahaemolyticus*	The LTBentero protein	CaMV 35S promoter/ *Agrobacterium tumefaciens*-mediated	5.29 µg/g FW	With complete Freund's adjuvant / incomplete Freund's adjuvant	Subcutaneously (10 mg)	Human	Induced IgG and mucosal IgA	[23]

FW: Fresh Weight, Tsb: Total Soluble Protein, NA: Not Available, FTW: Fresh Tuber Weight, CaMV: Cauliflower Mosaic Virus, AIDS: Acquired Immune Deficiency Syndrome. ST: Solanum tuberosum, LE: Lycopersicon esculentum.

Table 2.
Plant-based vaccines developed by nuclear transformation/stable expression system.

like *Physcomitrella patens* is able to perform post-translational modifications like N-glucosylation identical to higher plants [17].

Considerable efforts have been made to develop methods that can simplify the purification procedure as the downstream processing of recombinant vaccine antigens will significantly increase production costs. ELPylation as one of these methods, generally increases the accumulation of transgenic proteins in plants as reported in various examples [18]. In addition, ELPylated proteins can be rapidly purified by membrane-based inverse transition cycling (mITC) procedure. Using ELPylation, in which an elastin-like polypeptide (ELP) consisting of a series of pentapeptides is fused to the end of the target protein, an increase in antigen accumulation is observed in both transiently and stably transformed plants [26]. It has also been reported that the immune response increases significantly with the proper folding and trimerization of the antigen. In the study, in which the hemagglutinin of the avian influenza virus (AIV HA) was produced as a monomer and ELPylated trimer form in the tobacco plant (transient in *Nicotiana benthamiana* and stable in *Nicotiana tabacum*), it was observed that the trimeric AIV HA form increased the specific immune response to HA compared to the monomeric form [26].

Virus-like particles that do not contain viral nucleic acids and are formed by viral capsid proteins become prominent as much more reliable candidates when compared with attenuated viral vaccines. Studies in which capsid proteins are expressed in plants by nuclear transformation have been shown to induce specific antibodies comparable to attenuated vaccines when administered orally with an adjuvant [27]. VP2, VP6 and VP7 capsid proteins of group A rotavirus, one of the most common cause of human pathogen, acute infantile and pediatric gastroenteritis worldwide, have been expressed in transgenic tobacco plants. It has been reported to be significantly higher IgG antibodies in the serum of mice of the group VP 2/6/7 (immunized with transgenic tobacco plants) than group VP 2/6 (immunized with transgenic tobacco plants) and the group RV (immunized with orally attenuated RV vaccine) group IgG antibodies are significant. In addition, it was reported that the serum IgG titer in the VP 2/6 group was almost as high as that found in the RV group [27].

Using the nuclear transformation system, plant-based vaccines can be developed on the basis of low-cost multiepitopic recombinant proteins to contain epitope variants that can induce broad-spectrum antibodies. In addition, sequences containing adjuvant activity can be added to these multiepitopic proteins. Multiepitopic vaccines produced in plants trigger local and systemic immune responses more than their counterparts produced in bacteria. In orally inoculated BALBc mice, lettuce-derived HIV C4 (V3) 6 multiepitopic protein showed a higher immunogenic potential than *E. coli*-derived HIV C4 (V3) 6 multiepitopic protein [28].

Plant systems are also used in the production of vaccines against various parasitic diseases other than bacteria and viruses. *Taenia solium* cysticercosis is one of the important parasitic diseases affecting human health especially in developing countries. Production of a low-cost and multi-epitope vaccine against this disease was achieved in the tobacco (*Nicotiana tabacum*) plants by *Agrobacterium*-mediated transformation. The developed transgenic lines were self-pollinated and T1 generation transgenic plants were obtained from the harvested seeds. In these plants, several vaccine-related antigens were expressed in a polyprotein system based on the ribosomal skip mechanism via the 2A sequence derived from the foot-and mouth virus, which induces self-cleavage events at the translational level [29].

2.2 Chloroplast transformation

This section emphasized on recent advances in creation and development of transplastomic plants to produce plant-derived vaccines and protection against infectious diseases. Plastid transformation has come to be advantageous for the production of vaccines due to the high copy number in a single transformation allowing high levels of transgene expression and the absence of effects leading to gene silencing and the lack of concerns about positional and pleiotropic effects. In addition, as a result of maternal inheritance, the containment of foreign genes in the chloroplast genome and the absence of transgenes in the pollen are other advantages. Moreover, chloroplast transformation ensures expression of multiple genes in prokaryotic-like operon systems. However, this system is limited by the insufficient number of the target plant variety and the trials of very few plant varieties according to nuclear transformation. Another disadvantage is the lack of glucosylation ability of chloroplasts. Therefore, difficulty of expression of eukaryotic human or viral genes in prokaryotic chloroplasts is most important barrier to the use of transplastomic plants. Some major costs associated with the production of recombinant proteins in fermentation-based systems can be reduced by using chloroplasts as bioreactors. Chloroplast-derived therapeutics, especially when administered orally, eliminate expensive purification steps, cold storage, transportation and sterile injection requirements [30].

Advances in transgenic approaches and the development of gene gun and biolistic (particle bombardment method) technologies have enabled the transgene to be transferred directly to living cells. With these systems where tungsten and gold particles are used as microcarriers, the transformation of plastids can be carried out effectively. As an alternative to this system, plastids are transformed via polyethylene glycol (PEG) [31, 32]. PEG-mediated transformation allows the simultaneous transformation of many samples as a simple and efficient method and enables a large number of transformed cells with a high survival rate. However, it has a lower success rate than biolistic-mediated transformation. Despite its high transformation efficiency, biolistic is not available in many laboratories and standardization of its protocol is very difficult. It has been shown repeatedly that recombinant proteins capable of eliciting protective mammalian immune responses can also be produced in chloroplasts of plants. Thanks to chloroplast transformation, vaccines can be produced against viral diseases such as polio (poliovirus), human immunodeficiency (HIV), human papilloma virus (HPV), as well as numerous contagious and fatal bacterial infections and diseases such as cholera, tuberculosis, plague and anthrax (**Table 3**).

Virus-like particles formed by viral capsid proteins that do not contain viral nucleic acids are frequently preferred within vaccine production in plants due to their ability to elicit protective immune responses. In this context, Lenzi et al. [56] reported that the self-assembled L1 capsid protein of human papilloma virus (HPV), that is the causative agent of cervical cancer which is one of the most common causes of death for women, can be produced in chloroplasts of the *Nicotiana tabacum* plant. Thus, in tobacco chloroplasts, the HPV-16 L1 vaccine could be produced by expression of the L1 protein from a natural viral ($L1^{v}$ gene) or a synthetic sequence ($L1^{pt}$ gene) optimized for expression, under the control of plastid expression signals. In addition, accumulation of L1 antigen was obtained only when the first 14 amino acids of the N-terminal domain of the ATPase beta subunit or the Rubisco large subunit were translationally fused to the N-terminal of the L1 protein.

The level of transgene expression in chloroplasts varies depending on the origin of the coding sequence. Although the amount of transcript depends on the high copy number of the transgenes, it is also closely related to the efficiency of the promoter chosen and the regulatory sequences that affect translation. Most of the transgenes expressed in chloroplasts utilize the psbA promoter. It has also been reported that the 5'UTR sequence of psbA shows higher translation activity compared to many 5'UTR sequences. For this reason, studies connected with improving codon optimization benefited from psbA sequences, and it has been reported that

Antigens	Disease/Pathogen	% of Total Soluble Protein or μg of antigen/g fresh biomass	Plant	Immunization	Transfer Method	Ref.
Viral protein 1	Poliovirus	—	*Nicotiana tabacum*	Orally (ORV)	Biolistic	[33]
gp120 and gp41 multiepitope protein	HIV	16 μg /g fresh weight	*Nicotiana tabacum*	Orally (ORV)	Biolistic	[34]
gp120 and gp41 multiepitope protein	HIV	16 μg /g fresh weight	*Nicotiana tabacum*	Orally (ORV)	Biolistic	[15]
ESAT-6 and Mtb72F	*Mycobacterium tuberculosis*	7.5% for ESAT-6 and 1.2% for Mtb72F	*Nicotiana tabacum*	Orally (ORV)	Biolistic	[35]
mmpI with Lymphotoxin-beta (LTB)	*Mycobacterium leprae and Mycobacterium avium*	—	*Nicotiana benthamiana*	—	Polyethylene glycol (PEG)	[36]
C4V3 polypeptide	HIV	—	*Nicotiana tabacum*	Orally (ORV)	Biolistic	[37]
L1 capsid protein fused with LTB	Human papillomavirus	2%	*Nicotiana tabacum*	—	Biolistic	[38]
dengue-3 serotype polyprotein (prM/E)	Dengue virus	—	*Lactuca sativa*	Orally (ORV)	Biolistic	[39]
cholera toxin-B AMA1 MSP1	*Vibrio cholera* (Cholera)-Plasmodium (Malaria)	13.17 and 10.11%	*Nicotiana tabacum*	Subcutaneously (SQV) or orally (ORV)	Biolistic	[40]
cholera toxin-B AMA1 MSP1	*Vibrio cholera* (Cholera)-Plasmodium (Malaria)	7.3 and 6.1%	*Lactuca sativa*	Subcutaneously (SQV) or orally (ORV)	Biolistic	[40]
D2 fibronectin-binding domain+CTB	*Staphylococcus aureus*	0.7%	*Chlamydomonas reinhardtii*	Orally (ORV)	Biolistic	[41]
VP1	Foot and mouth dis ease virus (FMDV)	51%	*Nicotiana tabacum*	Inoculated	Biolistic	[42]
HIV-1 $Pr55^{gag}$ polyprotein	HIV	6.75%	*Nicotiana benthamiana*	—	Biolistic	[43]
labile toxin B subunit heat-stable toxin (LTB-ST)	ETEC-induced diarrhoeal disease	2.3%	*Nicotiana tabacum*	Orally (ORV)	Biolistic	[44]

Antigens	Disease/Pathogen	% of Total Soluble Protein or µg of antigen/g fresh biomass	Plant	Immunization	Transfer Method	Ref.
E7 and E7CP HPV protein	Human papillomavirus	0.1% and 0.5%	*Nicotiana tabacum*	—	Biolistic	[45]
multi-epitope DPT fusion protein	*Corynebacterium diphtheria, Bordetella pertussis, Clostridium tetani*	0.8%	*Nicotiana tabacum*	Orally (ORV)	Biolistic	[46]
F1-V	*Yersinia pestis* (Plague)	14.8%	*Nicotiana tabacum*	Orally (ORV)	Biolistic	[47]
HIV-1 Pr55Gag, p24 and p17/p24	HIV	4 µg/g fresh weight	*N. benthamiana*	Intramuscularly	*A. tumefaciens* (chloroplast-targeted)	[48]
L1 capsid protein	Human papillomavirus	24%	*Nicotiana tabacum*	Intraperitoneal (IP)	Biolistic	[49]
L1 capsid protein	Human papillomavirus	1.5%	*Nicotiana tabacum*	—	Biolistic	[50]
Structural protein E2	Swine fever virus (CSFV)	1–2%	*Nicotiana tabacum*	Subcutaneous and intragastric	Biolistic	[51]
TetC	*Clostridium tetani* (Tetanus)	10%	*Nicotiana tabacum*	Oral and İntranasal (IN)	Biolistic	[52]
SARS-CoV Spike Protein	Severe acute respiratory syndrome coronavirus	0.2%	*Nicotiana tabacum*	Orally (ORV)	Biolistic	[53]
Protective antigen (PA)	*Bacillus anthracis* (Anthrax)	2.7 and 1.7%	*Nicotiana tabacum*	—	Biolistic	[54]
VP6	Rotavirus	3%	*Nicotiana tabacum*	—	Biolistic	[55]

Table 3.
Plant-based vaccines developed by chloroplast transformation.

codon optimization significantly increases translation in chloroplasts [57]. It has been indicated that the expression of the gene of the vaccine subunit is increased 50 times in chloroplasts by codon optimization [33].

In recent years, the production of vaccine antigens in the chloroplasts of edible plants like lettuce and ensuring oral vaccination have come to the fore as an important development in order to eliminate the economically demanding and extremely expensive steps such as fermentation, purification, cold storage, cold chain and transportation. Lyophilized plant cells stored at ambient temperature retain their efficacy and antigen folding/assembly, thus eliminating the need for cold chain [33]. Thus, chloroplast bioreactors have become an important alternative to the production of fermentation-based vaccine antigens. Arlen et al. [47] have achieved the production of high levels of F1-V antigen, as much as 14.8% of the total soluble protein. In a study conducted by aerosol challenge with *Y. pestis*, 33% of subcutaneously F1-V (with adjuvant) immunized mice survived, while 88% of orally F1-V immunized mice survived. These data presented that oral doses of vaccine antigens produced in chloroplasts may be effective in eliciting protective immune responses in vivo. In yet another study, an increase in IgG1 and IgA titers was reported with oral boosting of low-cost, cold chain-free, chloroplast made viral protein 1 (VP1) subunit vaccine [28]. By producing multiple vaccine antigens in chloroplasts, immunity can be improved against more than one infectious disease at the same time. Cholera toxin-B subunit (CTB) fused malarial vaccine antigens apical membrane antigen-1 (AMA1) and merozoite surface protein-1 (MSP1), expressed in lettuce and tobacco chloroplasts inhibited the proliferation of malaria in red blood cells by inducing antigen-specific antibodies in mice [40].

Besides terrestrial plant systems, photosynthetic unicellular alga *Chlamydomonas reinhardtii*'s chloroplasts are used for vaccine production against various pathogens. Algae vaccines can remain stable at room temperature for more than 1.5 years and show faster and more controllable growth characteristics than other members of the plant kingdom. D2 fibronectin-binding domain of CTB fused *Staphylococcus aureus* was stably expressed in *C. reinhardtii* algae and as a result of algae-based inoculation of mice, the amount of pathogen decreased and 80% of mice were protected against the lethal dose of *S. aureus* [41]. It was reported in another study that the fusion protein was developed against tuberculosis caused by *Mycobacterium tuberculosis*, one of the leading fatal diseases, can be stored in lyophilized leaf for up to 6 months at room temperature and preserves its stability and proper folding/assembly [36].

Production of vaccine antigens (HIV gag transgene; $Pr55^{gag}$) by biolistic-mediated chloroplast transformation resulted in significantly greater protein accumulation than *Agrobacterium*-mediated nuclear transformation which vaccine antigens postranslationally targeted into plastids [43]. In transient expression experiments, various cellular organelles (cytosol, apoplast, endoplasmic reticulum, chloroplast and mitochondrion) were targeted. It was reported that specific $Pr55^{gag}$ sequences were expressed only in chloroplasts. The synthetic gene encoding the HIV C4V3 recombinant protein known to induce both systemic and mucosal immune responses in mammalian systems was expressed in chloroplasts of *Nicotiana tabacum* plants. It has been reported that the obtained plant-derived C4V3 elicits systemic and mucosal antibody responses in BALB /c mice by oral immunization [37]. Multepitopic protein (Multi-HIV) derived from HIV gp120 and gp41 envelope proteins expressed in tobacco chloroplasts also induced immune responses and T-helper specific responses by oral immunization in BALB/c mice [34].

2.3 Plant virus based expression system

Plant viruses are generally described as safe for humans and animals. Therefore, they are preferred for the production of therapeutic molecules. TMV (tobacco mosaic virus), PVX (potato virus X), BaMV (bamboo mosaic virus), CPMV (cowpea mosaic virus) are highly stable to high temperature, pressure and pH conditions and can be purified from host plants in amounts exceeding hundreds of mg/kg plant biomass [58]. Virus-like particles (VLPs) are multi-subunit molecules consist of self-assembly protein structures that are the same or highly similar to the general structure of authentic virus. Due to the fact that VLPs do not contain viral nucleic acids, therefore conversion to infectious viruses is not possible which is an important risk factor for attenuated vaccines. In recent years, the use of both VLPs and plant viruses for vaccine production has increased rapidly (**Table 4**). There are various recombinant vaccine production strategies that stand out in which different vaccine antigens can be produced by using plant viral structures. Launch vector-based on virus for plant transient expression system, plant virus-based vector systems by the incorporation of 2A peptide, plant virus conjugated with purified antigen from bacterial expression system are some of the systems by which recombinant vaccine antigens can be produced.

Multiple doses of multivalent vaccine can be administered in the immunization of mice without any adverse effects. In addition, multivalent subunit vaccines can be developed against various diseases using an efficient TMV-based delivery platform. A multivalent subunit vaccine consisting of the combination of OmpA, DnaK chaperone and Tul4 protective antigens of the *Francisella tularensis* pathogen bacterium has been reported to be safe. *F. tularensis* proteins were chemically conjugated to the TMV surface and the developed subunit vaccine strongly induced humoral immune response [60]. Chen et al. [59] produced a candidate vaccine (BJ2A CVP) in *Chenopodium quinoa* and *N. benthamiana* against Japanese encephalitis virus (JEV) using a bamboo mosaic virus-based chimeric virus particle (CVP) strategy. *Chenopodium quinoa* plant is preferred to reduce the side effects of nicotine and other alkaloids found in tobacco species in the purification of BJ2A CVP.

The genomes of both RNA and DNA viruses can be modified for recombinant protein production. Geminiviral replicon systems are one of the plant-viral based expression systems used to increase the expression of vaccine antigens in plants. In geminiviral derived vectors with a single stranded DNA genome, the viral genes encoding the coat and movement proteins are deleted and the expression cassette for protein of interest is inserted. In these strategies, it has been reported that the viral vector has transient expression only 4 days after it was transferred to *Nicotiana benthamiana* plant leaves by *Agrobacterium* infiltration and maintained this expression level up to 7 days [70]. On the other hand, by using plants with low secondary metabolite content such as lettuce in geminiviral replicon systems based on bean yellow dwarf virus, virus-like particles could be produced at high expression levels. Thus, vaccine candidates such as Norwalk virus capsid protein-VLP (NVCP-VLP) can be purified from lettuce plants without losing their functional activity. Plant virus-based expression methods along with *Agrobacterium*-mediated agroinfiltration are often preferred for increasing low expressions of vaccine antigens in plants and improving the feasibility of plant-based vaccines. The expression of recombinant proteins were carried out in *Nicotiana benthamiana* plant by using the Launch vector-based plant transient expression system with agroinfiltration, and it was reported that vaccine candidates caused up to 100% protection against diseases that could be used for bioterrorism such as anthrax after purification [63]. In another study, it has been reported that Consensus domain III of dengue virus E glycoprotein (cEDIII) shows high expression with plant virus-based expression (5.2 mg/g dry weight of leaf tissues) against Dengue virus [61]. Launch vector technologies, which are used in the production of VLP-based recombinant vaccines especially against infectious diseases, are also used in the production of protective vaccines such as malaria transmission blocking vaccines (TBVs). TBVs prevent successful

Antigens	Disease/ Pathogen	Recombinant vaccine platforms for production of antigen	Plant	Immunization	Virus	Ref.
JEV envelope protein domain III (EDIII)	Japanese encephalitis virus (JEV)	BaMV-based vector system	*Chenopodium quinoa*	Intraperitoneally (IP)	Bamboo mosaic virus (BaMV)	[59]
OmpA-like protein (OmpA), chaperone protein DnaK and lipoprotein Tul4	*Francisella tularensisis* (Tularemia)	Bacterial system (*E. coli*) and chemically conjugated	*Nicotiana benthamiana* for TMV-Lysine production)	Intranasally (IN), subcutaneously (SC.)	Tobacco mosaic virus (TMV)	[60]
Dengue virus envelope glycoprotein (E) domain III	Dengue virus	TMV-based vector system	*Nicotiana benthamiana* (Agroinfiltration)	Orally (ORV)	Tobacco Mosaic Virus (TMV)	[61]
RhoA-derived peptide (Antiviral peptide production/inhibitor)	Respiratory syncytial virus (RSV)	TMV-based vector system	*Nicotiana benthamiana* (Agroinfiltration)	—	Tobacco Mosaic Virus (TMV)	[62]
Protective antigen (PA)	*Bacillus anthracis*	Launch vector-based plant transient expression system	*Nicotiana benthamiana* (Agroinfiltration)	Intramuscularly (IM)	Tobacco mosaic virus (TMV)	[63]
Pfs25 VLP	*Plasmodium falciparum* (Malaria)	Launch vector-based on Tobacco mosaic virus	*Nicotiana benthamiana*	Intramuscularly (IM)	Tobacco Mosaic Virus (TMV)	[64]
Norwalk virus capsid Protein (NVCP)	Norwalk virus	BeYDV-based geminiviral replicon system	*Lactuca sativa* (Agroinfiltration)	—	Bean yellow dwarf virus (BeYDV)	[65]
PCV2 capsid protein	Porcine Circovirus (PCV)	CMV-based expression system	*Nicotiana tabacum* (inoculated with purified virions)	Intraperitoneally (IP)	Cucumber mosaic virus (CMV)	[66]
Extracellular domain M2 protein (M2e) fused to hepatitis B core antigen (HBc)	Influenza	Potato X virus-based vector system	*Nicotiana benthamiana* (Agroinfiltration)	Intraperitoneally (IM)	Potato X virus (PXV)	[67]

Antigens	Disease/ Pathogen	Recombinant vaccine platforms for production of antigen	Plant	Immunization	Virus	Ref.
Hemagglutinin (HA) protein	H5N1 influenza virus	Launch vector-based plant transient expression system	*Nicotiana benthamiana* (Agroinfiltration)	Subcutaneously (s.c.)	Tobacco mosaic virus (TMV)	[68]
BTV coat proteins	Bluetongue virus (BTV)	Cowpea mosaic virus–based HyperTrans (CPMV-HT) and associated pEAQ plant transient expression vector system	*Nicotiana benthamiana*	Subcutaneously (s.c.)	Cowpea mosaic virus	[69]

Table 4.
Plant-based vaccines developed by plant virus-based expression system.

sporogonic development of the sexual stage *Plasmodium falciparum* parasites ingested by female *Anopheles* mosquitoes [64]. Thus, the spread of parasites in endemic populations by transmission from human to mosquito is prevented [41]. Another plant virus-based expression approach is the fusion of the vaccine antigen to the virus-like carrier particle, either genetically or by chemical cross-linking [67]. It has been reported that the potato X virus-based recombinant viral vector provides a high level of expression of the hybrid protein consisting of influenza virus M2 protein (M2e) fused to hepatitis B core antigen (HBc) in *Nicotiana benthamiana* plants. This vector was transferred to plant leaves by agroinfiltration and the hybrid protein was synthesized to 1–2% of the total soluble protein.

2.4 Transient expression

In addition to stable transgene expression (nuclear or plastid transformation), transient expression is often preferred for the expression of genes encoding vaccine proteins in plant tissues. Since transient expression does not contain chromosomal integration, it is not affected by position effect. In addition, the expression of extrachromosomal transgenes can be detected 3 hours after transfer and can last for about 10 days [14]. The major advantage of transient expression is production of vaccine antigen rapidly at low cost and high yield. In addition, the easy applicability of the system and its ability to allow the production of complex proteins composed of subunits encourages its use against the novel viral diseases that emerge suddenly. Plant-based vaccines developed by transient expression are given in **Table 5**.

Margolin [73] achieved *Agrobacterium*-mediated transient expression of soluble HIV Env gb140 antigens in *Nicotiana benthamiana* plant. It has been reported that rabbits immunized intramuscularly with lectin affinity purified antigens developed binding antibodies and neutralizing antibodies at high titers. The production of nucleo capsid (N) and membrane protein (M), two important antigens of SARS-CoV, was achieved in *Nicotiana benthamiana* plant with transient expression created by using both virus-based vector and agroinfiltration. In addition, tobacco leaves infiltrated with *Agrobacterium tumafaciens* C58 and GV3101 strains harboring the pBI-Mvector containing the M protein, SARS-CoV M protein production was successfully achieved without using any post-transcriptional gene silencing suppressors [78].

It is imperative to produce vaccines at rates which offset mutation frequency of viral infections such as influenza, for which a new and unique epidemic strain appears within a few years. In recent years, recombinant vaccines become prominent as one of the most important options to solve this problem. New recombinant strategies provided by plant biotechnology and the production of plant-based vaccines are becoming widespread in the struggle with pandemic and epidemic diseases such as Influenza A H1N1, Influenza H5N1, plague, Ebola, Zika, SARS-CoV and SARS-Cov-2 [78–81]. Plant-based vaccines developed for pandemic and epidemic diseases are given in **Table 6**.

Especially the commercial scale production of these vaccines and their examination at Phase I, Phase II and Phase III levels beyond the functional evaluations in animal models indicates that in the future, plant-derived vaccines will be an important part of the struggle against pandemic and epidemic diseases. Plant-based vaccines produced in commercial scale and candidate vaccines are given in **Table 7** by companies. For instance, COVID-19 (severe acute respiratory syndrome coronavirus 2/SARS-CoV-2), which has become a major threat to global health, has also significantly impacted the world economy and social mobility. So far, with its high contagiousness, rapid spreading nature and high mortality rate, more than 2.5 million people have died from COVID-19, and more than 116 million people have

Plant	Pathogen/ Disease	Antigent	Transfer Method	% of Total Soluble Protein or μg of antigen/g fresh biomass	Adjuvant	Immunization	Target organism	Immun response	Ref.
Nicotiana benthamiana	HPV-16/ Cervical cancer	E7 protein fused 16E7SH	*Agrobacterium tumefaciens*-mediated	0.4–6 g/kg FW	With adjuvant/ (Freund's incomplete adjuvant)	Subcutaneous/ 5 μg	Human	Tumor size decrease and IgG induction	[71]
Nicotiana benthamiana	Hepatitis B virus	HBsAg	MagnICON viral Vectors/ *Agrobacterium tumefaciens*-mediated	0,64 mg/g FW	With adjuvant (alum)	Intraperitoneal 346 mIU/mL at week	Human	anti-HBsAg response	[72]
Nicotiana benthamiana	HIV/AIDS	Subtype C Envelope gp140	*Agrobacterium tumefaciens*-mediated	4.9–6.2 mg/kg FW	With adjuvant (Alhydrogel®)	Intramuscularly/ 50 μg	Human	NA	[73]
Nicotiana benthamiana	Poliovirus (PV) type 3	Capsid protein VP1	*Agrobacterium tumefaciens*-mediated	60 mg/kg FW	—	Intraperitoneal / intramuscular	Human	—	[74]
Nicotiana benthamiana	Infectious Bursal Disease Virus (IBDV)	Structural VP2 protein	*Agrobacterium tumefaciens*-mediated	1% TSB	With adjuvant (Freund's complete adjuvant and Freund's incomplete adjuvant)	Intramuscular/ (12 μg of VP2)	Poultry	NA	[75]
Nicotiana benthamiana	FMDV	Capsid precursor P1-2A and the protease 3C fusion	*Agrobacterium tumefaciens*-mediated	3–4 mg/kg FW	With adjuvant (Montanide ISA 50)	Intraperitoneal/ 500 ng	meat-producing animals	NA	[76]

Plant	Pathogen/ Disease	Antigent	Transfer Method	% of Total Soluble Protein or μg of antigen/g fresh biomass	Adjuvant	Immunization	Target organism	Immun response	Ref.
Nicotiana benthamiana	*Plasmodium falciparum*/ Malaria	Plasmodium Surface Protein Pf38 fused to Red floresan Protein	CaMV 35S promoter/ *Agrobacterium tumefaciens*-mediated	4–12 μg/g FW	With Gerbu MM Adjuvant	Intraperitoneal (17 μg)	Human	Induced IgG	[77]

FW: Fresh Weight, Tsb: Total Soluble Protein, NA: Not Available, CaMV: Cauliflower Mosaic Virus.

Table 5.
Plant-based vaccines developed by transient expression system.

Spreading area	Plant	Pathogen/ Disease	Antigent	Transfer Method	% of Total Soluble Protein or µg of antigen/g fresh biomass	Adjuvant	Immunization	Immun response	Ref.
E	*Oryza sativa*	*Vibrio cholerae*/ *Cholera*	Cholera toxin B-subunit	*Agrobacterium tumefaciens* mediated	NA	Witout adjuvant	Orally/150 mg seed	Induced IgG and mucosal IgA	[82]
E	*Nicotiana benthamiana*	SARS-CoV	SARS-CoV nucleocapsid (rN) protein	*Agrobacterium tumefaciens* mediated	0.8–1% of the TSP	With complete Freund's adjuvant/ incomplete Freund's adjuvant	Intraperitoneal/500 mg fresh tobacco leaves	Induced IgG1 and IgG2/ Increased IFN-_ and IL-10/not changed IL-2 and IL-4	[83]
P	*Nicotiana benthamiana*	Influenza A H1N1	Soluble protein H1/ H1-VLP	*Agrobacterium tumefaciens* mediated	NA	-	-	Induced CD4+ and CD8+ T cells	[84]
E	*Nicotiana benthamiana*	Influenza A H5N1	Matrix protein 2 ectodomain (M2e) fused to N-terminal proline-rich domain (Zera®) of the γ-zein protein of maize	*Agrobacterium tumefaciens* mediated	125–205 mg/ kg FW	Without adjuvant	Intramuscular/4.5 µg	Induced IgG	[85]
E	*Nicotiana tabacum*/*Lycopersicon esculentum*	*Yersinia pestis*/Plague	Major capsular protein F1-V antigen fused	*Agrobacterium tumefaciens* mediated	NT: 1-4% FW LE: 4-10% mg DW	With adjuvant NT: aluminum hydroxide t/LE: cholera toxin	NT: Subcutaneously (10µg purified)/LE: Orally (2 g tomato fruit)	Induced serum IgG1, IgG2a and mucosal IgA	[86]

DOI: http://dx.doi.org/10.5772/intechopen.97861

Spreading area	Plant	Pathogen/ Disease	Antigent	Transfer Method	% of Total Soluble Protein or µg of antigen/g fresh biomass	Adjuvant	Immunization	Immun response	Ref.
E	*Nicotiana benthamiana*	Flavivirus/ Yellow fever (YF)	YF virus envelope protein (YFE) fusion to the bacterial enzyme lichenase (YFE-LicKM)	*Agrobacterium tumefaciens* mediated	NA	With Alhydrogel adjuvant	Intramuscularly 5 µg x3/5 µg x 2/30 µg x 3	Induced IgG, İncreased IFNγ	[87]
E	*Nicotiana tabacum*	Ebola virus (EBOV)	Envelope-associated protein VP40	*Agrobacterium tumefaciens* mediated	2.6 µg/g FW	With complete Freund's adjuvant/ incomplete Freund's adjuvant	Orally (25 ng)/ Subcutaneously (125 ng)	Induced IgM, IgG and intestinal IgA	[88]
E	*Nicotiana benthamiana*	Zika virus (ZIKV)	Envelope (E) protein	*Agrobacterium tumefaciens* mediated	160 µg/g FW	With aluminium hydroxide gel adjjuvant	Subcutaneously/50 µg x 24	Induced IgG1 and IgG2, Increased IFN-γ, IL-4 and IL-6	[89]
P	*Nicotiana benthamiana*	SARS-CoV-2	Spike specific monoclonal antibody (mAb) CR3022	*Agrobacterium tumefaciens* mediated	130 µg/g FW	-	-	-	[90]

P: Pandemic, E: Epidemic, FW: Fresh Weight, Tsb: Total Soluble Protein, NA: Not Available, DW: Dry Weight. NT: Nicotiana tabacum, LE: Lycopersicon esculentum.

Table 6.
Plant-based vaccines developed for pandemic and epidemic diseases.

iBio, Inc			[91]
IBIO-201 Prophylaxis of SARS-CoV-2 spike protein fused lichenase protein / *Nicotiana benthamiana*	IBIO-400 Prophylaxis of Classical swine fever (CSF) CSFV E2 glycoprotein / *Nicotiana benthamiana*		
Medicago Inc.			[92]
MT-7529 Prophylaxis of H7N9 influenza / Phase 2 / *Nicotiana benthamiana*	MT-2355 Prophylaxis of pertussis, diphtheria, tetanus, poliomyelitis and prophylaxis of Hib infection in infants / Phase 3 / *Nicotiana benthamiana*	MT-2271 Prophylaxis of seasonal influenza / *Nicotiana benthamiana*	
MT-5625 Prophylaxis of rotavirus gastroenteritis / Phase 1 / *Nicotiana benthamiana*	MT-2766 Prophylaxis of SARS-CoV-2/ Phase 2 / *Nicotiana benthamiana*	MT-8972 Prophylaxis of H5N1 influenza / Phase 2 / *Nicotiana benthamiana*	
Icon Genetics			[93, 94]
ZMapp Prophylaxis of Ebola virus / *Nicotiana benthamiana*	Norovirus Vaccine Prophylaxis of Norovirus / Phase I / *Nicotiana benthamiana*		
SoyMeds, Inc.			[95]
soy-mSEB Prophylaxis of Staphylococcal Enterotoxin B / *Glycine max*			
Fraunhofer CMB			[96]
Prophylaxis of H1N1 influenza	Prophylaxis of H1N1 influenza	Prophylaxis of Malaria	

Table 7.
Plant-based vaccines producing companies and commercial products.

been infected with COVID-19 worldwide [97]. In order to control the pandemic, the whole world work hard to develop new strategies to be applied in the field of health, to deliver vital medical supplies to those in need, to develop and apply safe and effective vaccines. Especially, the development of vaccines and drugs for this new pathogenic Coronavirus, which emerged suddenly and mutated at certain times, became an inevitable target. Until now, the number of vaccines in preclinical development are 182 and the number of vaccines in clinical development is 74, worldwide [8]. Plant-based vaccines have also proven that they can play an active role in fighting against COVID-19 with their promising results in preclinical and clinical stages. Along with the initiative of commercial companies using plant biotechnology, transient expression of the SARS-CoV-2 antigen in plants was achieved, and a plant-based COVID-19 vaccine candidate was produced with a high-scale production technology [81]. The COVID-19 vaccine developed by Medicago company using a plant-based platform started Phase II clinical trials. In this approach, virus-like particles (SARS-CoV-2 spike protein self-assambles into VLPs) could be produced by transient expression in *Nicotiana benthamiana* plants within just 20 days after the acquisition of SARS-CoV-2 [81]. For Phase I, vaccine administration in healthy adults was performed as two intramuscular doses and at 3 different dose levels (S protein content 3.75, 7.5 or 15 μg), either alone or with adjuvant (AS03 or CpG1018). After the second dose administration, the adjuvanted CoVLP was able to induce humoral and cellular responses for all dose levels. Another company,

BAT's US Bio-tech arm, Kentucky BioProcessing (KBP), announced that the COVID-19 vaccine candidate will be evaluated in the Phase I trial after the Food and Drug Administration (FDA) accepts the Investigational New Drug application. This candidate vaccine was developed using plant-based technology and it was stated that the active ingredient of the vaccine could be rapidly produced in a short period of 6 weeks according to conventional methods. In still another initiative, iBio, Inc. company announced that it was working on the subunit IBIO-201 candidate vaccine against SARS-CoV 2 immediately after the pandemic was declared. The candidate vaccine was produced by *Agrobacterium*-mediated infiltration in leaves of *Nicotiana benthamiana* plants and the target protein was purified and formulated for the final product. The candidate vaccine has been reported to induce the production of anti-spike neutralizing antibodies in immunized mice. Due to the new mutations of SARS-CoV-2, it has become important to produce subunits or virus-like particles as vaccine candidates at low cost-per-dose and higher production scale.

2.5 Evaluation of plant-based vaccines side effect

Vital part of vaccine research is the risk assessment through randomized, double-blind placebo-controlled multicentre trials. Vaccine side effects can be evaluated under two categories as common side effects (high fever, vomiting, dizziness, anxiety and nausea) and rare side effects (risk of hospitalization, death or long-term morbidity). Same evaluation processes is required for plant-based vaccines as all traditional and recombinant vaccines. In literature, there are various studies in emphasis to safety and side effects of plant-based vaccines.

Plant-based vaccines can be evaluated in two different ways: cases in which the plant content is directly applied in pure form and the cases where the vaccine content is isolated and mixed with adjuvant. Phase studies were initiated for many candidate vaccines, where antigen or VLP was produced and then mixed with adjuvant before injection. In reported Phase I vaccine case against influenza A presented common side effects on volunteers as high fever, vomiting, dizziness, anxiety and nausea [96]. Similarly, in another case local effects occurred in the vaccination area and 93% of side effects were mild effects [98]. McCormick et al., stated in their studies that volunteers showed symptoms that were described as severe at a very low rate, but recovered within 1–2 days without the need for medical intervention, and the vaccine candidate was quite safe [99]. Moreover, Pillet et al. tested the vaccines with 300 healthy adults and 450 volunteers over the age of 50 in their phase III study on two different age groups. As a result, a higher rate of fatigue was observed in volunteers over 50 years old [100]. Ward et al. reported that the most common side effect was pain at the injection site in their studies on 22,854 volunteers and mortality rate was slightly higher for inactivated.

vaccine comparing to virus like particles [101]. Chichester et al., (2018) reported that 94% of the volunteers developed at least one of the side effects of high fever, vomiting, dizziness, anxiety and nausea [102]. In all of these studies with both oral and injectable vaccines, the observed effects were evaluated as mild to moderate. In this way, plant-based production has been defined as an effective and safe vaccine production tool [103]. Production of the adjuvants to be used to stimulate mucosal and peripheral immunity in the plant or the selection of plant which produce appropriate secondary metabolites that can act as mucosal adjuvants contributes to the decrease in the incidence of side effects. The number of studies in which plant-based vaccine candidates have passed to phase studies were much less than classical vaccine studies. When these studies are evaluated, there is no significantly increased side effect risk report concerning plant-based vaccines against any other vaccines production options [104].

2.6 Legal regulations involving plant-made pharmaceuticals

Regulatory processes of the vaccine development, approval, authorization, licensing, distribution, and marketing are as challenging as the production. There are both national and centralized regulatory agencies. These agencies emphasize on scientific evaluation of data, quality of the product, safety for human use, verification of reported efficacy and authenticity of product labels. In USA, centralized regulatory agency is Food and Drug Administration known as FDA. As plant based-vaccines are considered in biological materials apart from chemical entities, plant-made pharmaceutics are regulated under Biologics License Application (BLA) [105]. The European Union (EU) members have both their own national regulation and the centralized regulation under European Medicines Agency (EMA) which acts as the counterpart of FDA in Europe. The Committee on Herbal Medicinal Products (HMPC) is the EMA's committee responsible for compiling and assessing scientific data on herbal substances, preparations and combinations, to support the harmonization of the European market [106]. However, considering the nature of plant based vaccines, they are under authorization of The Committee for Medical Products for Human Use (CHMP) which plays a vital role in the authorization of medicines in the European Union (EU).

Strictness of the regulation is mostly based on the production method and the host plant. Non-food plants as *Nicotiana* species and the controlled indoor production methods as plant tissue cultures or climate rooms are not regulated as strictly as in field production of GM food plants. However, there are opinion differences between EU and USA regarding the plant-made PMPs. In USA, regulation is at product level. Field use of GM plants and plants used in vaccine development are under secondary regulation of United States Department of Agriculture (USDA). In May, 2020 USDA Animal and Plant Health Inspection Service (APHIS) released revision on 7 Code of Federal Regulation (CFR) Part 340 regulations (85 Fed. Reg. 29790) which regulates the importation, interstate movement, and environmental release of genetically engineered organisms that are or may be plant pests [107]. Therefore, scrutiny on biotechnological production is further reduced in favor of researchers and companies. PMPs are currently under regulation of Title 21 (Food and Drugs) US CFR along with other traditionally produced counterparts. In EU regulation is on both production and final product level. All GM plant derived pharmaceuticals including plant based vaccines are under the same regulation of other biotechnology derived drugs which is indicated in European Commission Directive 2001/83/EC (on the Community code relating to medicinal products for human use) [108] and Regulation No 726/2004 (for the authorization and supervision of medicinal products for human and veterinary use) [109]. Also, if the host plant is a human food or animal feed source, cultivation and release of GM plant is under regulation of EC Directive 2001/18/EC (on the deliberate release into the environment of genetically modified organisms) [110] and 1829/2003/EC (on genetically modified food and feed).

Regulatory approval of a PMPs and other biotechnologically derived products may take up to a year depending on the legal response limits of the regulation agencies. However, Covid-19 pandemic triggered a realization on EU and UK for national and global needs. Therefore, product review, conditional approval and deployment timelines are significantly reduced for PMPs [111].

In conclusion, infectious diseases threatened humanity countless times throughout history. In particular, as pandemic and epidemic diseases killed millions of people, it increases in importance to develop safe and cost-effective vaccines and their storage and rapid distribution. Apart from many diseases such as diphtheria, cholera, typhoid, tuberculosis, which are controlled by vaccine campaigns in developed countries, new vaccine production systems using recombinant DNA technologies are needed for emerging diseases such as COVID-19, MERS-CoV, avian influenza, Ebola, Zika and possible future infections. Plant-based vaccine production for humans and animals stands out as an important alternative that can be used to overcome the disadvantages of existing conventional vaccines. Within the scope of plant biotechnology, it became possible to produce cost-effective, immunogenic and safe vaccines thanks to the development of gene transfer strategies to plants and improvements in amount, isolation and purification and addition of adjuvant for production of recombinant vaccine antigens in plants. It is an undeniable fact that the possibilities that recombinant vaccines can offer us will increase with new standards and legal regulations to be introduced for the development, approval, authorization, licensing, distribution and marketing of such vaccines. In scope of future preventive healthcare, it is hard to assume monopoly of one particular vaccine technology. There will always be some ups and downs in any vaccine production methods. However, plant based vaccines represent considerable strong suits and offer swift and viable solutions over traditional and other recombinant vaccines.

Author details

Sinan Meriç*, Tamer Gümüş and Alp Ayan
Department of Molecular Biology and Genetic, Faculty of Science and Letters, Istanbul Kultur University, Ataköy, Istanbul, Turkey

*Address all correspondence to: s.meric@iku.edu.tr

References

[1] J. Wilimowska, Ethnic diversity in the Ptolemaic Fayum, Acta Antiq. Hung. 56 (2016) 287-292. https://doi.org/10.1556/068.2016.56.3.1.

[2] G.R. Storey, The population of ancient Rome, Antiquity. 71 (1997) 966-978. https://doi.org/10.1017/S0003598X00085859.

[3] J.M.C. Pearce-Duvet, The origin of human pathogens: evaluating the role of agriculture and domestic animals in the evolution of human disease, Biol. Rev. 81 (2006) 369-382. https://doi.org/10.1017/S1464793106007020.

[4] WHO, World Health Organization, (2021). https://www.who.int/.

[5] G. Kumar, L. Karthik, K.V.B. Rao, Plant vaccines: An overview, in: Microb. Bioprospecting Sustain. Dev., Springer Singapore, 2018: pp. 249-250. https://doi.org/10.1007/978-981-13-0053-0_13.

[6] A.J. Pollard, E.M. Bijker, A guide to vaccinology: from basic principles to new developments, Nat. Rev. Immunol. 21 (2021) 83-100. https://doi.org/10.1038/s41577-020-00479-7.

[7] WHO vaccine-preventable diseases: monitoring system. 2020 global summary, (n.d.).

[8] WHO, COVID-19 vaccines, (2021). https://www.who.int/emergencies/diseases/novel-coronavirus-2019/covid-19-vaccines.

[9] W. Wang, Q. Wu, J. Yang, K. Dong, X. Chen, X. Bai, X. Chen, Z. Chen, C. Viboud, M. Ajelli, H. Yu, Global, regional, and national estimates of target population sizes for covid-19 vaccination: Descriptive study, BMJ. 371 (2020). https://doi.org/10.1136/bmj.m4704.

[10] T. Thanh Le, Z. Andreadakis, A. Kumar, R. Gómez Román, S. Tollefsen, M. Saville, S. Mayhew, The COVID-19 vaccine development landscape, Nat. Rev. Drug Discov. 19 (2020) 305-306. https://doi.org/10.1038/d41573-020-00073-5.

[11] J. Lang, S.C. Wood, Development of orphan vaccines: An industry perspective, Emerg. Infect. Dis. 5 (1999) 749-756. https://doi.org/10.3201/eid0506.990602.

[12] P.P. Pastoret, Human and animal vaccine contaminations, Biologicals. 38 (2010) 332-334. https://doi.org/10.1016/j.biologicals.2010.02.015.

[13] T. Cardi, *P. lenzi*, P. Maliga, Chloroplasts as expression platforms for plant-produced vaccines, Expert Rev. Vaccines. 9 (2010) 893-911. https://doi.org/10.1586/erv.10.78.

[14] T. V. Komarova, S. Baschieri, M. Donini, C. Marusic, E. Benvenuto, Y.L. Dorokhov, Transient expression systems for plant-derived biopharmaceuticals, Expert Rev. Vaccines. 9 (2010) 859-876. https://doi.org/10.1586/erv.10.85.

[15] S. Rosales-Mendoza, N. Rubio-Infante, E. Monreal-Escalante, D.O. Govea-Alonso, A.L. García-Hernández, J.A. Salazar-González, O. González-Ortega, L.M.T. Paz-Maldonado, L. Moreno-Fierros, Chloroplast expression of an HIV envelop-derived multiepitope protein: Towards a multivalent plant-based vaccine, Plant Cell. Tissue Organ Cult. 116 (2014) 111-123. https://doi.org/10.1007/s11240-013-0387-y.

[16] T. Pniewski, M. Milczarek, J. Wojas-Turek, E. Pajtasz-Piasecka, J. Wietrzyk, M. Czyż, Plant lyophilisate carrying S-HBsAg as an oral booster vaccine against HBV, Vaccine. 36 (2018) 6070-6076. https://doi.org/10.1016/j.vaccine.2018.09.006.

[17] L. Orellana-Escobedo, S. Rosales-Mendoza, A. Romero-Maldonado, J. Parsons, E.L. Decker, E. Monreal-Escalante, L. Moreno-Fierros, R. Reski, An Env-derived multi-epitope HIV chimeric protein produced in the moss Physcomitrella patens is immunogenic in mice, Plant Cell Rep. 34 (2015) 425-433. https://doi.org/10.1007/s00299-014-1720-6.

[18] L. Dedieu, D.M. Floss, M. Mockey, G. Zanello, D. Brosson, M. Diogon, R. Frutos, T. Bruel, V. Rodrigues, E. Garzon, C. Chevaleyre, M. Berri, H. Salmon, U. Conrad, Expression and immunogenicity of the mycobacterial Ag85B/ESAT-6 antigens produced in transgenic plants by elastin-like peptide fusion strategy, J. Biomed. Biotechnol. 2010 (2010). https://doi.org/10.1155/2010/274346.

[19] D.E. Webster, S.D. Smith, R.J. Pickering, R.A. Strugnell, I.B. Dry, S.L. Wesselingh, Measles virus hemagglutinin protein expressed in transgenic lettuce induces neutralising antibodies in mice following mucosal vaccination, Vaccine. 24 (2006) 3538-3544. https://doi.org/10.1016/j.vaccine.2006.02.002.

[20] J.Y. Zhou, L.Q. Cheng, X.J. Zheng, J.X. Wu, S. Bin Shang, J.Y. Wang, J.G. Chen, Generation of the transgenic potato expressing full-length spike protein of infectious bronchitis virus, J. Biotechnol. 111 (2004) 121-130. https://doi.org/10.1016/j.jbiotec.2004.03.012.

[21] X. Zhang, N.A. Buehner, A.M. Hutson, M.K. Estes, H.S. Mason, Tomato is a highly effective vehicle for expression and oral immunization with Norwalk virus capsid protein, Plant Biotechnol. J. 4 (2006) 419-432. https://doi.org/10.1111/j.1467-7652.2006.00191.x.

[22] E. Loza-Rubio, E. Rojas-Anaya, J. López, M.T. Olivera-Flores, M. Gómez-Lim, G. Tapia-Pérez, Induction of a protective immune response to rabies virus in sheep after oral immunization with transgenic maize, expressing the rabies virus glycoprotein, Vaccine. 30 (2012) 5551-5556. https://doi.org/10.1016/j.vaccine.2012.06.039.

[23] E. Trujillo, S. Rosales-Mendoza, C. Angulo, A multi-epitope plant-made chimeric protein (LTBentero) targeting common enteric pathogens is immunogenic in mice, Plant Mol. Biol. 102 (2020) 159-169. https://doi.org/10.1007/s11103-019-00938-3.

[24] K. Kashima, Y. Yuki, M. Mejima, S. Kurokawa, Y. Suzuki, S. Minakawa, N. Takeyama, Y. Fukuyama, T. Azegami, T. Tanimoto, M. Kuroda, M. Tamura, Y. Gomi, H. Kiyono, Good manufacturing practices production of a purification-free oral cholera vaccine expressed in transgenic rice plants, Plant Cell Rep. 35 (2016) 667-679. https://doi.org/10.1007/s00299-015-1911-9.

[25] Y. Yuki, M. Nojima, O. Hosono, H. Tanaka, Y. Kimura, T. Satoh, S. Imoto, S. Uematsu, S. Kurokawa, K. Kashima, M. Mejima, R. Ouchida-Nakahashi, Y. Uchida, T. Marui, K. Fujihashi, N. Yoshikawa, F. Nagamura, H. Kiyono, Needle- and Cold Chain-Free Rice-Based Oral Vaccine, MucoRice-CTB Shows Microbiota-Dependent Immunogenicity in Humans, SSRN Electron. J. (2020). https://doi.org/10.2139/ssrn.3487818.

[26] H.T. Phan, J. Pohl, D.M. Floss, F. Rabenstein, J. Veits, B.T. Le, H.H. Chu, G. Hause, T. Mettenleiter, U. Conrad, ELPylated haemagglutinins produced in tobacco plants induce potentially neutralizing antibodies against H5N1 viruses in mice, Plant Biotechnol. J. 11 (2013) 582-593. https://doi.org/10.1111/pbi.12049.

[27] Y.M. Yang, X. Li, H. Yang, Y. Qian, Y. Zhang, R.X. Fang, X.Y. Chen, Immunogenicity and virus-like particle formation of rotavirus capsid proteins

produced in transgenic plants, Sci. China Life Sci. 54 (2011) 82-89. https://doi.org/10.1007/s11427-010-4104-3.

[28] D.O. Govea-Alonso, N. Rubio-Infante, A.L. García-Hernández, J.T. Varona-Santos, S.S. Korban, L. Moreno-Fierros, S. Rosales-Mendoza, Immunogenic properties of a lettuce-derived C4(V3)6 multiepitopic HIV protein, Planta. 238 (2013) 785-792. https://doi.org/10.1007/s00425-013-1932-y.

[29] E. Monreal-Escalante, B. Bañuelos-Hernández, M. Hernández, G. Fragoso, T. Garate, E. Sciutto, S. Rosales-Mendoza, Expression of Multiple Taenia Solium Immunogens in Plant Cells Through a Ribosomal Skip Mechanism, Mol. Biotechnol. 57 (2015) 635-643. https://doi.org/10.1007/s12033-015-9853-6.

[30] D. Verma, N.P. Samson, V. Koya, H. Daniell, A protocol for expression of foreign genes in chloroplasts, Nat. Protoc. 3 (2008) 739-758. https://doi.org/10.1038/nprot.2007.522.

[31] C. O'Neill, G. V. Horváth, É. Horváth, P.J. Dix, P. Medgyesy, Chloroplast transformation in plants: polyethylene glycol (PEG) treatment of protoplasts is an alternative to biolistic delivery systems, Plant J. 3 (1993) 729-738. https://doi.org/10.1111/j.1365-313X.1993.00729.x.

[32] A.H. Díaz, H.U. Koop, Nicotiana tabacum : PEG-mediated plastid transformation, Methods Mol. Biol. 1132 (2014) 165-175. https://doi.org/10.1007/978-1-62703-995-6_9.

[33] H. Chan, Y. Xiao, W.C. Weldon, S.M. Oberste, K. Chumakov, H. Daniell, Cold chain and virus-free chloroplast-made booster vaccine to confer immunity against different poliovirus serotypes, Plant Biotechnol. J. 14 (2016) 2190-2200. https://doi.org/10.1111/pbi.12575.

[34] N. Rubio-Infante, D.O. Govea-Alonso, A. Romero-Maldonado, A.L. García-Hernández, D. Ilhuicatzi-Alvarado, J.A. Salazar-González, S.S. Korban, S. Rosales-Mendoza, L. Moreno-Fierros, A Plant-Derived Multi-HIV Antigen Induces Broad Immune Responses in Orally Immunized Mice, Mol. Biotechnol. 57 (2015) 662-674. https://doi.org/10.1007/s12033-015-9856-3.

[35] P.S. Lakshmi, D. Verma, X. Yang, B. Lloyd, H. Daniell, Low Cost Tuberculosis Vaccine Antigens in Capsules: Expression in Chloroplasts, Bio-Encapsulation, Stability and Functional Evaluation In Vitro, PLoS One. 8 (2013) e54708. https://doi.org/10.1371/journal.pone.0054708.

[36] S.W. Hassan, Towards Generation of Transplastomic Tobacco, Expressing a 35 kdaProtein as an Antigen: A Step towards Affordable Plant made Vaccine against Mycobacterium, Cloning Transgenes. 03 (2013) 1. https://doi.org/10.4172/2168-9849.1000117.

[37] N. Rubio-Infante, D.O. Govea-Alonso, Á.G. Alpuche-Solís, A.L. García-Hernández, R.E. Soria-Guerra, L.M.T. Paz-Maldonado, D. Ilhuicatzi-Alvarado, J.T. Varona-Santos, L. Verdín-Terán, S.S. Korban, L. Moreno-Fierros, S. Rosales-Mendoza, A chloroplast-derived C4V3 polypeptide from the human immunodeficiency virus (HIV) is orally immunogenic in mice, Plant Mol. Biol. 78 (2012) 337-349. https://doi.org/10.1007/s11103-011-9870-1.

[38] M.T. Waheed, N. Thönes, M. Müller, S.W. Hassan, J. Gottschamel, E. Lössl, H.-P. Kaul, A.G. Lössl, Plastid expression of a double-pentameric vaccine candidate containing human papillomavirus-16 L1 antigen fused with LTB as adjuvant: transplastomic plants show pleiotropic phenotypes, Plant Biotechnol. J. 9 (2011) 651-660. https://doi.org/10.1111/j.1467-7652.2011.00612.x.

[39] A.P. Kanagaraj, D. Verma, H. Daniell, Expression of dengue-3 premembrane and envelope polyprotein in lettuce chloroplasts, Plant Mol. Biol. 76 (2011) 323-333. https://doi.org/10.1007/s11103-011-9766-0.

[40] A. Davoodi-Semiromi, M. Schreiber, S. Nalapalli, D. Verma, N.D. Singh, R.K. Banks, D. Chakrabarti, H. Daniell, Chloroplast-derived vaccine antigens confer dual immunity against cholera and malaria by oral or injectable delivery, Plant Biotechnol. J. 8 (2010) 223-242. https://doi.org/10.1111/j.1467-7652.2009.00479.x.

[41] I.A.J. Dreesen, G.C. El Hamri, M. Fussenegger, Heat-stable oral alga-based vaccine protects mice from Staphylococcus aureus infection, J. Biotechnol. 145 (2010) 273-280. https://doi.org/10.1016/j.jbiotec.2009.12.006.

[42] E.M. Lentz, M.E. Segretin, M.M. Morgenfeld, S.A. Wirth, M.J.D. Santos, M.V. Mozgovoj, A. Wigdorovitz, F.F. Bravo-Almonacid, High expression level of a foot and mouth disease virus epitope in tobacco transplastomic plants, Planta. 231 (2010) 387-395. https://doi.org/10.1007/s00425-009-1058-4.

[43] *N. Scotti*, F. Alagna, E. Ferraiolo, G. Formisano, L. Sannino, L. Buonaguro, A. De Stradis, A. Vitale, L. Monti, S. Grillo, F.M. Buonaguro, T. Cardi, High-level expression of the HIV-1 Pr55gag polyprotein in transgenic tobacco chloroplasts, Planta. 229 (2009) 1109-1122. https://doi.org/10.1007/s00425-009-0898-2.

[44] S. Rosales-Mendoza, Á.G. Alpuche-Solís, R.E. Soria-Guerra, L. Moreno-Fierros, L. Martínez-González, A. Herrera-Díaz, S.S. Korban, Expression of an *Escherichia coli* antigenic fusion protein comprising the heat labile toxin B subunit and the heat stable toxin, and its assembly as a functional oligomer in transplastomic tobacco plants, Plant J. 57 (2009) 45-54. https://doi.org/10.1111/j.1365-313X.2008.03666.x.

[45] M. Morgenfeld, M.E. Segretin, S. Wirth, E. Lentz, A. Zelada, A. Mentaberry, L. Gissmann, F. Bravo-Almonacid, Potato Virus X Coat Protein Fusion to Human Papillomavirus 16 E7 Oncoprotein Enhance Antigen Stability and Accumulation in Tobacco Chloroplast, Mol. Biotechnol. 43 (2009) 243-249. https://doi.org/10.1007/s12033-009-9195-3.

[46] R.E. Soria-Guerra, A.G. Alpuche-Solís, S. Rosales-Mendoza, L. Moreno-Fierros, E.M. Bendik, L. Martínez-González, S.S. Korban, Expression of a multi-epitope DPT fusion protein in transplastomic tobacco plants retains both antigenicity and immunogenicity of all three components of the functional oligomer, Planta. 229 (2009) 1293-1302. https://doi.org/10.1007/s00425-009-0918-2.

[47] P.A. Arlen, M. Singleton, J.J. Adamovicz, Y. Ding, A. Davoodi-Semiromi, H. Daniell, Effective plague vaccination via oral delivery of plant cells expressing F1-V antigens in chloroplasts, Infect. Immun. 76 (2008) 3640-3650. https://doi.org/10.1128/IAI.00050-08.

[48] A. Meyers, E. Chakauya, E. Shephard, F.L. Tanzer, J. Maclean, A. Lynch, A.L. Williamson, E.P. Rybicki, Expression of HIV-1 antigens in plants as potential subunit vaccines, BMC Biotechnol. 8 (2008) 53. https://doi.org/10.1186/1472-6750-8-53.

[49] A. Fernández-San Millán, S.M. Ortigosa, S. Hervás-Stubbs, P. Corral-Martínez, J.M. Seguí-Simarro, J. Gaétan, P. Coursaget, J. Veramendi, Human papillomavirus L1 protein expressed in tobacco chloroplasts self-assembles into virus-like particles that are highly immunogenic, Plant Biotechnol. J. 6 (2008) 427-441. https://doi.org/10.1111/j.1467-7652.2008.00338.x.

[50] *P. lenzi*, *N. Scotti*, F. Alagna, M.L. Tornesello, A. Pompa, A. Vitale, A. De Stradis, L. Monti, S. Grillo, F.M. Buonaguro, P. Maliga, T. Cardi, Translational fusion of chloroplast-expressed human papillomavirus type 16 L1 capsid protein enhances antigen accumulation in transplastomic tobacco, Transgenic Res. 17 (2008) 1091-1102. https://doi.org/10.1007/s11248-008-9186-3.

[51] H.B. Shao, D.M. He, K.X. Qian, G.F. Shen, Z.L. Su, The expression of classical swine fever virus structural protein E2 gene in tobacco chloroplasts for applying chloroplasts as bioreactors, Comptes Rendus - Biol. 331 (2008) 179-184. https://doi.org/10.1016/j.crvi.2007.12.007.

[52] J.S. Tregoning, S. Clare, F. Bowe, L. Edwards, N. Fairweather, O. Qazi, P.J. Nixon, P. Maliga, G. Dougan, T. Hussell, Protection against tetanus toxin using a plant-based vaccine, Eur. J. Immunol. 35 (2005) 1320-1326. https://doi.org/10.1002/eji.200425453.

[53] H.Y. Li, S. Ramalingam, M.L. Chye, Accumulation of recombinant SARS-CoV spike protein in plant cytosol and chloroplasts indicate potential for development of plant-derived oral vaccines, Exp. Biol. Med. 231 (2006) 1346-1352. https://doi.org/10.1177/153537020623100808.

[54] J. Watson, V. Koya, S.H. Leppla, H. Daniell, Expression of Bacillus anthracis protective antigen in transgenic chloroplasts of tobacco, a non-food/feed crop, Vaccine. 22 (2004) 4374-4384. https://doi.org/10.1016/j.vaccine.2004.01.069.

[55] I. Birch-Machin, C.A. Newell, J.M. Hibberd, J.C. Gray, Accumulation of rotavirus VP6 protein in chloroplasts of transplastomic tobacco is limited by protein stability, Plant Biotechnol. J. 2 (2004) 261-270. https://doi.org/10.1111/j.1467-7652.2004.00072.x.

[56] P.L. Ae, *N. Scotti*, A.E. Fiammetta, A. Ae, M.L. Tornesello, A.E. Andrea, P. Ae, A.V. Ae, A. De, S. Ae, L.M. Ae, S. Grillo, A.E. Franco, M. Buonaguro, A.E. Pal, M. Ae, T. Cardi, Translational fusion of chloroplast-expressed human papillomavirus type 16 L1 capsid protein enhances antigen accumulation in transplastomic tobacco, Springer. (n.d.). https://doi.org/10.1007/s11248-008-9186-3.

[57] K.C. Kwon, H.T. Chan, I.R. León, R. Williams-Carrier, A. Barkan, H. Daniell, Codon optimization to enhance expression yields insights into chloroplast translation, Plant Physiol. 172 (2016) 62-77. https://doi.org/10.1104/pp.16.00981.

[58] I. Balke, A. Zeltins, Use of plant viruses and virus-like particles for the creation of novel vaccines, Adv. Drug Deliv. Rev. 145 (2019) 119-129. https://doi.org/10.1016/j.addr.2018.08.007.

[59] T.-H. Chen, C.-C. Hu, J.-T. Liao, Y.-L. Lee, Y.-W. Huang, N.-S. Lin, Y.-L. Lin, Y.-H. Hsu, Production of Japanese Encephalitis Virus Antigens in Plants Using Bamboo Mosaic Virus-Based Vector, Front. Microbiol. 8 (2017) 788. https://doi.org/10.3389/fmicb.2017.00788.

[60] S. Banik, A.A. Mansour, R.V. Suresh, S. Wykoff-Clary, M. Malik, A.A. McCormick, C.S. Bakshi, Development of a Multivalent Subunit Vaccine against Tularemia Using Tobacco Mosaic Virus (TMV) Based Delivery System, PLoS One. 10 (2015) e0130858. https://doi.org/10.1371/journal.pone.0130858.

[61] M.Y. Kim, Y.S. Jang, M.S. Yang, T.G. Kim, High expression of consensus dengue virus envelope glycoprotein domain III using a viral expression system in tobacco, Plant Cell. Tissue Organ Cult. 122 (2015) 445-451. https://doi.org/10.1007/s11240-015-0781-8.

[62] B. Ortega-Berlanga, K. Musiychuk, Y. Shoji, J.A. Chichester, V. Yusibov, O.

Patiño-Rodríguez, D.E. Noyola, Á.G. Alpuche-Solís, Engineering and expression of a RhoA peptide against respiratory syncytial virus infection in plants, Planta. 243 (2016) 451-458. https://doi.org/10.1007/s00425-015-2416-z.

[63] J.A. Chichester, S.D. Manceva, A. Rhee, M. V. Coffin, K. Musiychuk, V. Mett, M. Shamloul, J. Norikane, S.J. Streatfield, V. Yusibov, A plant-produced protective antigen vaccine confers protection in rabbits against a lethal aerosolized challenge with bacillus anthracis ames spores, in: Hum. Vaccines Immunother., Landes Bioscience, 2013: pp. 544-552. https://doi.org/10.4161/hv.23233.

[64] R.M. Jones, J.A. Chichester, V. Mett, J. Jaje, S. Tottey, S. Manceva, L.J. Casta, S.K. Gibbs, K. Musiychuk, M. Shamloul, J. Norikane, V. Mett, S.J. Streatfield, M. van de Vegte-Bolmer, W. Roeffen, R.W. Sauerwein, V. Yusibov, A Plant-Produced Pfs25 VLP Malaria Vaccine Candidate Induces Persistent Transmission Blocking Antibodies against Plasmodium falciparum in Immunized Mice, PLoS One. 8 (2013) e79538. https://doi.org/10.1371/journal.pone.0079538.

[65] H. Lai, Q. Chen, Bioprocessing of plant-derived virus-like particles of Norwalk virus capsid protein under current Good Manufacture Practice regulations, Plant Cell Rep. 31 (2012) 573-584. https://doi.org/10.1007/s00299-011-1196-6.

[66] Á. Gellért, K. Salánki, K. Tombácz, T. Tuboly, E. Balázs, A Cucumber Mosaic Virus Based Expression System for the Production of Porcine Circovirus Specific Vaccines, PLoS One. 7 (2012) e52688. https://doi.org/10.1371/journal.pone.0052688.

[67] N. V. Ravin, R.Y. Kotlyarov, E.S. Mardanova, V. V. Kuprianov, A.I. Migunov, L.A. Stepanova, L.M. Tsybalova, O.I. Kiselev, K.G. Skryabin, Plant-produced recombinant influenza vaccine based on virus-like HBc particles carrying an extracellular domain of M2 protein, Biochem. 77 (2012) 33-40. https://doi.org/10.1134/S000629791201004X.

[68] Y. Shoji, H. Bi, K. Musiychuk, A. Rhee, A. Horsey, G. Roy, B. Green, M. Shamloul, C.E. Farrance, B. Taggart, N. Mytle, N. Ugulava, S. Rabindran, V. Mett, J.A. Chichester, V. Yusibov, Plant-derived hemagglutinin protects ferrets against challenge infection with the A/Indonesia/05/05 strain of avian influenza, Vaccine. 27 (2009) 1087-1092. https://doi.org/10.1016/j.vaccine.2008.11.108.

[69] E.C. Thuenemann, A.E. Meyers, J. Verwey, E.P. Rybicki, G.P. Lomonossoff, A method for rapid production of heteromultimeric protein complexes in plants: assembly of protective bluetongue virus-like particles, Plant Biotechnol. J. 11 (2013) 839-846. https://doi.org/10.1111/pbi.12076.

[70] Q. Chen, J. He, W. Phoolcharoen, H.S. Mason, Geminiviral vectors based on bean yellow dwarf virus for production of vaccine antigens and monoclonal antibodies in plants, Hum. Vaccin. 7 (2011) 331-338. https://doi.org/10.4161/hv.7.3.14262.

[71] M. Whitehead, P. Öhlschläger, F.N. Almajhdi, L. Alloza, P. Marzábal, A.E. Meyers, I.I. Hitzeroth, E.P. Rybicki, Human papillomavirus (HPV) type 16 E7 protein bodies cause tumour regression in mice, BMC Cancer. 14 (2014). https://doi.org/10.1186/1471-2407-14-367.

[72] Z. Huang, K. LePore, G. Elkin, Y. Thanavala, H.S. Mason, High-yield rapid production of hepatitis B surface antigen in plant leaf by a viral expression system, Plant Biotechnol. J. 6 (2008) 202-209. https://doi.org/10.1111/j.1467-7652.2007.00316.x.

[73] E. Margolin, R. Chapman, A.E. Meyers, M.T. van Diepen, P. Ximba, T. Hermanus, C. Crowther, B. Weber, L. Morris, A.L. Williamson, E.P. Rybicki, Production and Immunogenicity of Soluble Plant-Produced HIV-1 Subtype C Envelope gp140 Immunogens, Front. Plant Sci. 10 (2019) 1378. https://doi.org/10.3389/fpls.2019.01378.

[74] J. Marsian, H. Fox, M.W. Bahar, A. Kotecha, E.E. Fry, D.I. Stuart, A.J. Macadam, D.J. Rowlands, G.P. Lomonossoff, Plant-made polio type 3 stabilized VLPs-A candidate synthetic polio vaccine, Nat. Commun. 8 (2017). https://doi.org/10.1038/s41467-017-00090-w.

[75] E. Gómez, M.S. Lucero, S. Chimeno Zoth, J.M. Carballeda, M.J. Gravisaco, A. Berinstein, Transient expression of VP2 in Nicotiana benthamiana and its use as a plant-based vaccine against Infectious Bursal Disease Virus, Vaccine. 31 (2013) 2623-2627. https://doi.org/10.1016/j.vaccine.2013.03.064.

[76] V. Ruiz, J. Baztarrica, E.P. Rybicki, A.E. Meyers, A. Wigdorovitz, Minimally processed crude leaf extracts of Nicotiana benthamiana containing recombinant foot and mouth disease virus-like particles are immunogenic in mice, Biotechnol. Reports. 20 (2018). https://doi.org/10.1016/j.btre.2018.e00283.

[77] T. Feller, P. Thom, N. Koch, H. Spiegel, O. Addai-Mensah, R. Fischer, A. Reimann, G. Pradel, R. Fendel, S. Schillberg, M. Scheuermayer, H. Schinkel, Plant-based production of recombinant plasmodium surface protein Pf38 and evaluation of its potential as a vaccine candidate, PLoS One. 8 (2013). https://doi.org/10.1371/journal.pone.0079920.

[78] O.C. Demurtas, *S. Massa*, E. Illiano, D. De Martinis, P.K.S. Chan, P. Di Bonito, R. Franconi, Antigen Production in Plant to Tackle Infectious Diseases Flare Up: The Case of SARS, Front. Plant Sci. 7 (2016) 54. https://doi.org/10.3389/fpls.2016.00054.

[79] N. Landry, S. Pillet, D. Favre, J.F. Poulin, S. Trépanier, B. Yassine-Diab, B.J. Ward, Influenza virus-like particle vaccines made in Nicotiana benthamiana elicit durable, poly-functional and cross-reactive T cell responses to influenza HA antigens, Clin. Immunol. 154 (2014) 164-177. https://doi.org/10.1016/j.clim.2014.08.003.

[80] F. Le Mauff, G. Mercier, *P. Chan*, C. Burel, D. Vaudry, M. Bardor, L.-P. Vézina, M. Couture, P. Lerouge, N. Landry, Biochemical composition of haemagglutinin-based influenza virus-like particle vaccine produced by transient expression in tobacco plants, Plant Biotechnol. J. 13 (2015) 717-725. https://doi.org/10.1111/pbi.12301.

[81] B.J. Ward, P. Gobeil, A. Séguin, J. Atkins, I. Boulay, P.Y. Charbonneau, M. Couture, M.A. D'Aoust, J. Dhaliwall, C. Finkle, K. Hager, A. Mahmood, A. Makarkov, M. Cheng, S. Pillet, P. Schimke, S. St-Martin, S. Trépanier, N. Landry, Phase 1 trial of a Candidate Recombinant Virus-Like Particle Vaccine for Covid-19 Disease Produced in Plants, MedRxiv. (2020) 2020.11.04.20226282. https://doi.org/10.1101/2020.11.04.20226282.

[82] K. Kashima, Y. Yuki, M. Mejima, S. Kurokawa, Y. Suzuki, • Satomi Minakawa, • Natsumi Takeyama, Y. Fukuyama, T. Azegami, T. Tanimoto, • Masaharu Kuroda, M. Tamura, Y. Gomi, • Hiroshi Kiyono, Good manufacturing practices production of a purification-free oral cholera vaccine expressed in transgenic rice plants, Springer. (1911). https://doi.org/10.1007/s00299-015-1911-9.

[83] N. Zheng, R. Xia, C. Yang, B. Yin, Y. Li, C. Duan, L. Liang, H. Guo, Q. Xie, Boosted expression of the SARS-CoV

nucleocapsid protein in tobacco and its immunogenicity in mice, Vaccine. 27 (2009) 5001-5007. https://doi.org/10.1016/j.vaccine.2009.05.073.

[84] A.I. Makarkov, M. Golizeh, E. Ruiz-Lancheros, A.A. Gopal, I.N. Costas-Cancelas, S. Chierzi, S. Pillet, N. Charland, N. Landry, I. Rouiller, P.W. Wiseman, M. Ndao, B.J. Ward, Plant-derived virus-like particle vaccines drive cross-presentation of influenza A hemagglutinin peptides by human monocyte-derived macrophages, Npj Vaccines. 4 (2019). https://doi.org/10.1038/s41541-019-0111-y.

[85] S. Mbewana, E. Mortimer, F.F.P.G. Pêra, I.I. Hitzeroth, E.P. Rybicki, Production of H5N1 influenza virus matrix protein 2 ectodomain protein bodies in tobacco plants and in insect cells as a candidate universal influenza vaccine, Front. Bioeng. Biotechnol. 3 (2015) 1-8. https://doi.org/10.3389/fbioe.2015.00197.

[86] M.L. Alvarez, H.L. Pinyerd, J.D. Crisantes, M.M. Rigano, J. Pinkhasov, A.M. Walmsley, H.S. Mason, G.A. Cardineau, Plant-made subunit vaccine against pneumonic and bubonic plague is orally immunogenic in mice, Vaccine. 24 (2006) 2477-2490. https://doi.org/10.1016/j.vaccine.2005.12.057.

[87] S. Tottey, Y. Shoji, R.M. Jones, J.A. Chichester, B.J. Green, K. Musiychuk, H. Si, S.D. Manceva, A. Rhee, M. Shamloul, J. Norikane, R.C. Guimarães, E. Caride, A.N.M.R. Silva, M. Simões, P.C.C. Neves, R. Marchevsky, M.S. Freire, S.J. Streatfield, V. Yusibov, Plant-produced subunit vaccine candidates against yellow fever induce virus neutralizing antibodies and confer protection against viral challenge in animal models, Am. J. Trop. Med. Hyg. 98 (2018) 420-431. https://doi.org/10.4269/ajtmh.16-0293.

[88] E. Monreal-Escalante, A.A. Ramos-Vega, J.A. Salazar-González, B. Bañuelos-Hernández, C. Angulo, S. Rosales-Mendoza, Expression of the VP40 antigen from the Zaire ebolavirus in tobacco plants, Planta. 246 (2017) 123-132. https://doi.org/10.1007/s00425-017-2689-5.

[89] M. Yang, H. Sun, H. Lai, J. Hurtado, Q. Chen, Plant-produced Zika virus envelope protein elicits neutralizing immune responses that correlate with protective immunity against Zika virus in mice, Plant Biotechnol. J. 16 (2018) 572-580. https://doi.org/10.1111/pbi.12796.

[90] K. Rattanapisit, B. Shanmugaraj, S. Manopwisedjaroen, P.B. Purwono, K. Siriwattananon, N. Khorattanakulchai, O. Hanittinan, W. Boonyayothin, A. Thitithanyanont, D.R. Smith, W. Phoolcharoen, Rapid production of SARS-CoV-2 receptor binding domain (RBD) and spike specific monoclonal antibody CR3022 in Nicotiana benthamiana, Sci. Rep. 10 (2020) 1-11. https://doi.org/10.1038/s41598-020-74904-1.

[91] iBio, ibionic vaccines, (2021). http://www.ibioinc.com/vaccines/.

[92] Mitsubishi Tanabe Pharma, R&D News, (2021). https://www.mt-pharma.co.jp/e/develop/news/.

[93] Icon Genetics, ZMapp, an experimental treatment for Ebola disease, (2021). https://www.icongenetics.com/technology/usage-of-magnicon-technology/.

[94] Icon Genetics, Icon Genetics clinical development of its novel norovirus vaccine reaches milestone of complete dosing of the first cohort, (2021). https://www.icongenetics.com/icon-genetics-clinical-development-of-its-novel-norovirus-vaccine-reaches-milestone-of-complete-dosing-of-the-first-cohort/.

[95] T.K. Inskeep, C. Stahl, J. Odle, J. Oakes, L. Hudson, K.L. Bost, K.J. Piller, Oral Vaccine Formulations Stimulate

Mucosal and Systemic Antibody Responses against Staphylococcal Enterotoxin B in a Piglet Model, Clin. Vaccine Immunol. 17 (2010) 1163-1169. https://doi.org/10.1128/CVI.00078-10.

[96] J.F. Cummings, M.L. Guerrero, J.E. Moon, P. Waterman, R.K. Nielsen, S. Jefferson, F.L. Gross, K. Hancock, J.M. Katz, V. Yusibov, J.A. Chichester A., R.M. Jones, S.D. Manceva D., B.J. Green J., M. Stow, F. Miao, G. Moonsammy, S.J. Streatfield, Safety and immunogenicity of a plant-produced recombinant monomer hemagglutinin-based influenza vaccine derived from influenza A (H1N1)pdm09 virus: A Phase 1 dose-escalation study in healthy adults, Vaccine. 32 (2014) 2251-2259. https://doi.org/10.1016/j.vaccine.2013.10.017.

[97] WHO, Coronavirus (COVID-19) Dashboard, (2021). https://covid19.who.int/.

[98] P. Maurer, G.T. Jennings, J. Willers, F. Rohner, Y. Lindman, K. Roubicek, W.A. Renner, P. Müller, M.F. Bachmann, A therapeutic vaccine for nicotine dependence: Preclinical efficacy, and phase I safety and immunogenicity, Eur. J. Immunol. 35 (2005) 2031-2040. https://doi.org/10.1002/eji.200526285.

[99] A.A. McCormick, S. Reddy, S.J. Reinl, T.I. Cameron, D.K. Czerwinkski, F. Vojdani, K.M. Hanley, S.J. Garger, E.L. White, J. Novak, J. Barrett, R.B. Holtz, D. Tusé, R. Levy, Plant-produced idiotype vaccines for the treatment of non-Hodgkin's lymphoma: Safety and immunogenicity in a phase I clinical study, Proc. Natl. Acad. Sci. U. S. A. 105 (2008) 10131-10136. https://doi.org/10.1073/pnas.0803636105.

[100] S. Pillet, J. Couillard, S. Trépanier, J.F. Poulin, B. Yassine-Diab, B. Guy, B.J. Ward, N. Landry, Immunogenicity and safety of a quadrivalent plant-derived virus like particle influenza vaccine candidate—Two randomized Phase II clinical trials in 18 to 49 and 50 years old adults, PLoS One. 14 (2019). https://doi.org/10.1371/journal.pone.0216533.

[101] B.J. Ward, A. Makarkov, A. Séguin, S. Pillet, S. Trépanier, J. Dhaliwall, M.D. Libman, T. Vesikari, N. Landry, Efficacy, immunogenicity, and safety of a plant-derived, quadrivalent, virus-like particle influenza vaccine in adults (18-64 years) and older adults (≥65 years): two multicentre, randomised phase 3 trials, Lancet. 396 (2020) 1491-1503. https://doi.org/10.1016/S0140-6736(20)32014-6.

[102] J.A. Chichester, B.J. Green, R.M. Jones, Y. Shoji, K. Miura, C.A. Long, C.K. Lee, C.F. Ockenhouse, M.J. Morin, S.J. Streatfield, V. Yusibov, Safety and immunogenicity of a plant-produced Pfs25 virus-like particle as a transmission blocking vaccine against malaria: A Phase 1 dose-escalation study in healthy adults, Vaccine. 36 (2018) 5865-5871. https://doi.org/10.1016/j.vaccine.2018.08.033.

[103] Z.J. Guan, B. Guo, Y.L. Huo, Z.P. Guan, J.K. Dai, Y.H. Wei, Recent advances and safety issues of transgenic plant-derived vaccines, Appl. Microbiol. Biotechnol. 97 (2013) 2817-2840. https://doi.org/10.1007/s00253-012-4566-2.

[104] A. Sahoo, A.K. Mandal, K. Dwivedi, V. Kumar, A cross talk between the immunization and edible vaccine: Current challenges and future prospects, Life Sci. 261 (2020) 118343. https://doi.org/10.1016/j.lfs.2020.118343.

[105] FDA, Biologics License Applications (BLA) Process (CBER), (2021). https://www.fda.gov/vaccines-blood-biologics/development-approval-process-cber/biologics-license-applications-bla-process-cber.

[106] European Medicines Agency's (EMA), Committee on Herbal Medicinal Products (HMPC), (2021).

https://www.ema.europa.eu/en/committees/committee-herbal-medicinal-products-hmpc.

[107] Legal Information Institute, 7 CFR Chapter III - ANIMAL AND PLANT HEALTH INSPECTION SERVICE, DEPARTMENT OF AGRICULTURE, (2021). https://www.law.cornell.edu/cfr/text/7/chapter-III.

[108] DIRECTIVE 2001/83/EC OF THE EUROPEAN PARLIAMENT AND OF THE COUNCIL of 6 November 2001 on the Community code relating to medicinal products for human use, n.d.

[109] laying down Community procedures for the authorisation and supervision of medicinal products for human and veterinary use and establishing a European Medicines Agency (Text with EEA relevance), (2004).

[110] DIRECTIVE 2001/18/EC OF THE EUROPEAN PARLIAMENT AND OF THE COUNCIL of 12 March 2001 on the deliberate release into the environment of genetically modified organisms and repealing Council Directive 90/220/EEC, n.d.

[111] D. Tusé, S. Nandi, K.A. McDonald, J.F. Buyel, The Emergency Response Capacity of Plant-Based Biopharmaceutical Manufacturing-What It Is and What It Could Be, Front. Plant Sci. 11 (2020) 594019. https://doi.org/10.3389/fpls.2020.594019.

Chapter 5

microRNA Utilization as a Potential Tool for Stress Tolerance in Plants

Jyoti Rani

Abstract

This chapter describe the possibilities of MicroRNAs (miRNAs) in crop plants gene expression regulation in different metabolic pathways. Several current researches have shown different environmental stresses induce abnormal expression of miRNA, thus signifying that miRNAs may be an appropriate tool for genetical improvement in plant for stress tolerance. These miRNAs mainly control gene expression through translational inhibition. Generally, stress induce miRNAs-based inhibition of their target mRNAs, however, positive transcription factors accumulated and become more active after mRNA inhibition. Initially, researchers were mainly focused on miRNA identification, appropriate to specific or multiple environmental condition, expression profiling and recognize their roles in stress tolerance. Transformed miRNA expression studied in some plant species for better understanding of plant development and stress tolerance such as heavy metal, salinity, temperature, drought and nutrient deficiency. All these findings indicate that miRNAs act as a potential tool for genetic engineering and to enhance stress tolerance in crop plants.

Keywords: RNA, miRNA, siRNA, biotic and abiotic stresses, plant development

1. Introduction

Plants form an essential portion of the earth system and used by man as food, shelter, and a great source of medicine. Main threats to plant products and productivity are various human actions, biotic and abiotic stresses like soil toxicity, climate change, water stress, microbial pathogen, insects, herbivores *etc*. Global industrialization and increasing human population are two main factors that promote environmental changes and also enhance the demand for crop production. However, climate change and ever-increasing demand of plant products has the ability to modify the atmospheric properties and modify soils, which can make crop yield, development and growth more difficult. One of the important ways involves to enhance the yield and crop productivity is by using environmentally friendly plant protection measures [1]. Improvement in molecular biology and biotechnology identified the microRNAs role at post transcriptional level in controlling important secondary metabolites synthesis pathway [2]. This chapter gives a brief description on discovery and biogenesis of miRNA. On the basis of current research, it also describes the miRNA-based strategies used as potential tool for gene regulators in biofuel sources, beverages, cereals, fruits, fibers and economically important crops.

Plants are sessile organisms, obstinately face adverse environmental perturbations termed as abiotic stresses, most important being drought, soil salinity, extreme temperatures, and heavy metals [3]. Abiotic stresses have become a major challenge due to their widespread nature and the devastating impacts on plant growth, yields and the quality of plant produce. However, plants have developed intricate mechanisms for sensing and responding to environmental changes [4]. To turn on protective mechanisms, plants trigger a network of genetic regulations including altered expression of large proportion of genes by transcriptional and/or translational regulations [5].

2. miRNA biosynthesis and their conservation in plants

miRNA is ranges from 20 to 24 nucleotide base pair in length and non-coding RNA molecule [6]. MIR gene (miRNA genes) found in intergenic area of genome but some miRNA gene also found within intronic region in both sense or antisense direction. miRNA biomolecules are grouped in the genome and transcribed collectively as long polycistronic RNAs. miRNA synthesis occurs in the nucleus (**Figure 1**).

MIR genes transcribed into a specific primary transcript by the RNA polymerase II [6]. These are distinctive and specific primary transcripts modified at 5′ end by capping with a modified nucleotide and polyadenylated tail at the 3′ end with some adenosines [7]. The miRNAs are further modified by DICER-LIKE (DCL1), serrate (SE) and Hyponastic leaves 1 (HYL1) into precursors miRNA (pre-miRNAs). Pre-miRNA is a hairpin like structure. miRNA: miRNA duplexes produce in nucleus by further modification of pre-miRNAs by DCL1 [7].

S-adenosyl-L-methionine dependent RNA methyltransferase enzyme responsible for duplex methylation and HUA enhancer 1 at their 3′ end (HEN1) [8].

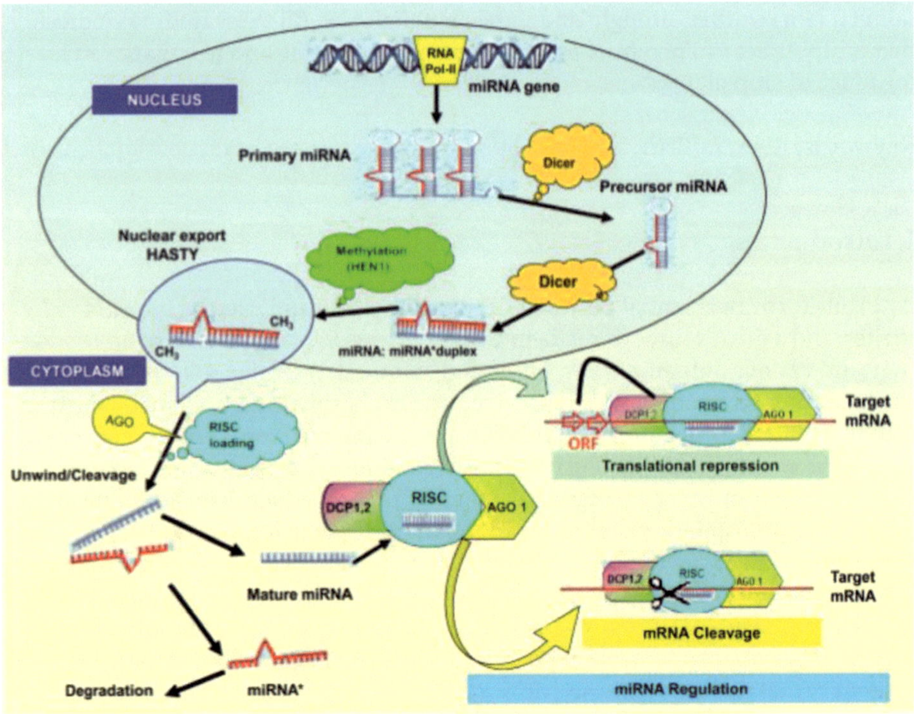

Figure 1.
Biosynthesis and mode of action of miRNA. Source: Djami-Tchatchou et al. [50].

Plant homologous of EXPORTIN 5 and HASTY transport this modified duplex into cytoplasm [9]. After transportation duplex separated by RNA-induced silencing complex (RISC) along with ARGONAUTE (AGO) protein [10, 11]. One strand of the miRNA is inserted into AGO protein comprised RISC complex, while other stand degraded by exosome along with small-RNA degrading nuclease [8]. This miRNA promotes the RISC loading to complementary mRNAs targets. Mature miRNAs showed resemblance with the target mRNA encourages site-specific cleavage of target mRNA while miRNAs along with some damaged base pairing with target mRNA leads inhibition of translation [8].

Some previous studies have shown that few miRNA bio-molecules are evolutionarily conserved among all significant plant lineages, including bryophytes, pteridophytes, gymnosperms, and angiosperms [12, 13]. miRNA families further divided into two different classes on the basis of miRNA diversification and conservation. The young miRNAs are expressed only in specific condition and at very low level although old miRNA is more evolutionarily conserved and highly expressed. In *Arabidopsis* some old studies reported the sign of regular synthesis and degradation of MIRNA genes. MIRNA gene producing loci either by aberrant transposition or replication/recombination from expressed gene sequences. Furthermore, it also showed that few miRNAs are lost during evolution [14].

3. Gene expression regulation by miRNA

For controlling various function of plant biology, specifically process control by transcription actors, miRNA diversity is significant [15]. miRNAs act as a significant controllers of gene expression and investigation on this aspect increasing now a days [2, 13]. Gene expression regulated by miRNA through high sequence similarity at the post-transcriptional level. Proper pairing between miRNA and targeted mRNA promotes the corresponding mRNA degradation and improper pairing between miRNA and target mRNA leads translation inhibition [13]. Poly(A) tail removal induced by mRNA which further promotes the destabilization and degradation of the target mRNAs [16]. Along with post-transcriptional gene expression regulation, miRNAs also decrease arbitrary fluctuation in transcript copy number and promote different metabolic pathways by transcription inhibition. miRNA of different length produces by different gene, as well as varied length miRNAs originated from the same gene. DCL1 enzyme mainly processed plant pre-miRNAs and produce 21 nucleotide base pair long mature miRNAs, but few other DCLs *i.e.* 2–4 can also be involved to produce miRNA of various lengths [17]. Diversity miRNA pool expanded by such miRNA heterogeneity and can efficiently enhance their monitoring possibilities. Additionally, miRNA diversity has practical used for the production of miRNA precursor-based expression casettes designed to produce artificial and sequence specific miRNAs [18].

4. Molecular techniques used for miRNA study

4.1 Techniques used for miRNA isolation, identification, and characterization

For miRNA identification first crucial step is to recognize their roles. Direct cloning, sequencing, genetic screening and some bioinformatic approaches commonly used for miRNAs identification [19]. In *Arabidopsis* first plant miRNAs identified such as miR156, miR159, miR164 and miR171 by isolating, cloning and sequencing of small RNA populations [20]. Small number of miRNAs have been recognized by genetic screening, mainly due to the redundancy and sequence similarity with other miRNA-coding genes. Genetic screening and activation-tagging approach used for miRNA identification and significant to separate prominent miRNA mutants like miR172a-2 [21]. JAW (jagged and wavy) loci also identified by genetic screening, which generate a microRNA (miR-JAW) that can leads the degradation of mRNAs of TCP genes which control leaf development [22]. Mutations like deletions/insertions/further promotes the loss/gain of miRNA binding sites during co-evolution of miRNAs and their target sites [23]. Mutation also contributes in identification for specific defect during development. In the earlier decade, thousands of plant miRNAs identified by both experimental and computational approaches. The main computational method used for miRNAs identification on the basis of sequence similarity against DNA/genome sequence of some important crop plants is BLAST (Basic Local Alignment Search Tool, www.ncbi.nlm.nih.gov/blast/) [24–27].

Direct cloning and genetic screening are experimentally old methods used for identification and functional characterization of miRNAs [20]. Sanger sequencing technique used after direct cloning for identification of sequence of base pair. But now a days next generation sequencing technology evolved as a powerful tool for discovery of novel miRNA and target identification in crop plants [9, 28]. Real time PCR and Northern bloting technique used for validation of identified miRNAs expression [29, 30]. Along with this, miRNA identification at protein level possible by using some other methods like mass spectrometry, proteins chromatography, protein foot printing, Western blotting, *etc*. at the protein level. Outcomes from these approaches showed that miRNAs act as rheostats to make fine-scale alterations in protein output. Further, miRNA identification promotes the development of different database which contains searchable evidence on the miRNAs. miRbase (http://www.mirbase.org) is the most significant and crucial bioinformatics tool used for miRNA research which is a searchable and comprehensive miRNA database mainly based on miRNA name, annotation, references and keyword [30]. A another bioinformatic database like The Plant MicroRNA Database (PMRD; http://mirnablog.com/plant-micrornadatabase-goes-online/) also contains information about plant miRNAs, like miRNA and their target(s), expression profiling, genome browser and secondary structure [31]. Several computational tools such as AthaMap (http://www.athamap.de/), CLC Genomic workbench 6 software (CLC Bio, Cambridge, MA, USA) and Next-Gen sequence databases also enhance the NGS performance along with the knowledge about miRNA and their role in Plants (https://mpss.udel.edu/index.php) [32].

4.2 Approaches for miRNA target screening and prediction

Several bioinformatics approaches and tools used for identification of miRNA target gene [33]. Sequence similarity scoring and secondary structure investigation are main bioinformatics criteria used for miRNA and target identification are: miRTarBase, miRTour (http://mirnablog.com/mirtour-plantmirna-and-target-prediction-tool/, psRNATarget (http://plantgrn. noble.org/psRNATarget/) and TAPIR (http://bioinformatics.psb.ugent.be/webtools/tapir/) [30]. Relationship between microRNA and its target confirmed by micRTarBase database (http://mirtarbase.mbc.nctu.edu.tw/) [34]. While the target mRNA expression levels can be observed by real-time PCR, for mapping the target site 5′-Rapid Amplification of cDNA Ends (RACE) used. Now a days, the degradome sequencing technique was developed for identification of comparative profusion of cleaved targets [30].

5. Functional roles of miRNAs

Several previous studies on model plants like *Arabidopsis* and some other plants confirmed that the miRNAs are involved in different biological pathways in which they play significant role in growth and development, upkeep of genome integrity, response to various biotic and abiotic stresses, signal transduction, homeostasis and hormone signaling pathways [35]. Different aspects of the miRNA regulatory roles in development, biosynthetic pathway and adaptive response to stresses have been reviewed [36] and more continues to be reported. Current study revealed that the miRNAs can be used to reprogram various cellular pathways, followed by the formation of microbe-associated molecular pattern (MAMP) molecules during pathogen attack, promote some dynamic changes in microtranscriptome along with differential transcriptional regulation in support of immunity and resistance [2].

6. miRNA-based strategies for improving plant crops

miRNA functional analysis confirmed their essential role in different biological and metabolic pathways in economically important plants [37]. Several studies indicated that miRNAs are act as a riboregulatory which control gene expression during plant growth and development, and response to biotic and abiotic stress. Consequently, miRNA-based genetic engineering technology is one of the most important tools which can play an essential role to enhance agricultural production in order to generate superior crop cultivars [38]. miRNA-based regulation of gene expression manipulated by several transgenic methods like overexpression of miRNA resistant gene, the production of artificial targets [37] is used. When miRNA of interest possesses a negative control on stress factors that miRNA will be an outstanding way for crop improvement, in which transgenic plants overexpressing the semi RNAs are susceptible to stresses [39].

Additional approach in which artificial miRNAs (amiRNAs) designed and used to suppress protein-coding mRNA expression. This is an advance gene silencing technique at post-transcriptional level which has been used efficiently in different plant species [40]. The artificial miRNAs technology was used to suppress the expression of the cucumber mosaic virus suppressor 2b [18]. After efficiently inhibiting 2b expression it also enhances resistance to transgenic tobacco plants against this virus [41]. This method was found an essential way to generate advance transgenic plants with high yield and improving crop tolerance to biotic and abiotic stresses [42].

6.1 Improve crop tolerance to abiotic and biotic stresses using microRNA-based approaches

Plant through molecular pathway replies to abiotic stresses involve interaction and interlinkage of different biosynthetic pathways involving gene expression regulation by miRNA and miRNA regulation [43]. Therefore, novel plant varieties produced with better environmental stress tolerance is essential for increasing crop productivity and quality. Previous studies proved that in tomato, the overexpression of miR169 increased water stress tolerance by inhibiting stomatal opening, which decreasing transpiration rate and water loss [44]. Salt stress tolerance in rice improved by decreasing expression of osa-MIR396c [45]. Similarly, transgenic rice lines produce by increasing the expression of miR398-resistant miRNA, due to this transgenic rice with more Cu- or Zn superoxide dismutases enzyme exhibited more tolerance to high water and salinity stress [46]. Induced expression of miR319 improve cold stress tolerance and also modify leaf morphology in rice [47].

Modernisation and advancement in miRNA research have also contributed in biotic stress tolerance in several plant species. Some previous research investigated that the Osa-miR7696 miRNA overexpression produce blast resistance transgenic rice [48]. In several plant species it was studied that the overexpression of miR393 effectively hinders the microbial growth and providing a disease resistant tool [3]. Few investigations showed that the miR160a overexpression positively regulate callose deposition induced by MAMP, while miR773 and miR398b negatively control MAMP-induced callose deposition and provide specific protection to bacterial infection [49]. Numerous studies conducted on model plants like Arabidopsis to investigate the miRNA role in plants [30], but till now only few researches were done on the significant role of miRNAs. Therefore, it is concluded after detailed analysis of old literature that miRNAs and its regulation play a crucial role in stress tolerance in plants.

6.2 microRNA-based approaches to improve plant growth and development

miRNAs play a crucial role in plant growth and development pathways such as leaf morphogenesis, apical dominance and plant biomass. Several miRNAs based new method used for production of transgenic plants for improving growth and development like plant morphology, fruit quality improvement, grain yield and more shelf life [50]. Overexpression of miR319 caused continuous growth of leaf margins and larger leaflets in tomato [51]. Previous studies in *Arabidopsis* indicated that the miR156 overexpression results in the increase in number of leaves, shape and size which can be 10 times higher than normal wild-type [52]. Recent investigation on switchgrass showed that the overexpression of miR156 repressed apical dominance which results into the increase in biomass and number of tiller by 58–100% in genetic modified plants [53]. Correspondingly, in tomato the more expression of miR156, increased the number of branching and leaves, and further enhance the plant biomass but supress the apical dominance [54].

Overexpression of OsmiR397 microRNA enhance rice productivity by increasing number of panicle branching and grain size [55]. Overexpression of miR319 in rice also increases the number of small and longitudinal veins [47]. In rice overstimulation of miR390 miRNA increasing the lateral root formation [56] by decreasing the gene expression of several lateral root growth repressors such as ARF2, ARF3, and ARF4. In *Medicago truncatula*, it was studeid that the overexpression of miR160, regulate the expression of gene which were significant for gravitropic movement and root development and induce several defects in root growth, root apical meristem organization and root nodule formation [57]. It has also been well-documented that miRNAs also play an essential role in controlling transition from vegetative to floral meristem in few crop plants such as *Arabidopsis*, maize and rice. Glossy 15 (APETALA-like gene) in maize mainly control the transition from vegetative to reproductive phase along with the leaf morphogenesis. An additional study on maize demonstrated that the miR172 overexpression leads to the inhibition of glossy15 gene expression which delayed phase transition from vegetative to reproductive [58]. Furthermore, [59] demonstrates that overexpression of miR172 and miR156 promotes the adult reproductive phase however low miR172 and high miR156 expression promote juvenility. Till now various role of miRNAs are studied in economically important crop plants either by increasing or decreasing miRNA expression, the manipulation of miRNA expression can be used for confirmation of miRNA functions and provide an effective way for improving plant growth, development, fruit, and seed development as well as plant biomass and productivity [50].

6.3 miRNA manipulation by genome editing technologies

To access the diverse roles of miRNAs in crops, a number of investigations conducted on miRNAs [45]. Although, due to small size of miRNA, the effectiveness of the well-known approaches for functional loss and inhibiting miRNA expression are comparatively less strong. Mainly, two genome editing techniques with modified nuclease enzyme are significant for selecting appropriate genome alteration [60]. These techniques mainly depend on the production of double-stranded breaks at target sequences by TALENs (Transcription Activator-Like Effector Nucleases) and CRISPR (Clustered, Regularly Interspaced Short Palindromic Repeats) DNA modification methods [61]. CRISPR-Cas9 technology developed as an advance RNA-dependent gene and genome editing tool due to its suitability to a variety of organisms [62]. A previous study on rice demonstrated that Cas9 also guided by modified gRNA for appropriate cut and genome editing of some selected crops [63]. One more study in soybean showed that the CRISPR/Cas9 system is very effective for removing a green fluorescent tagged protein transgene and modifying nine different endogenous loci [64].

CRISPR/Cas9 genome editing technology specifically and strongly decreases the 96% expression of miRNAs [62]. This method has transformed gene editing abilities and has been useful in several model plants such as *Arabidopsis*, tobacco *etc.* and other crops plant such as rice, wheat, maize, tomato and sorghum. But non-coding RNA editing by CRISPR/Cas method in plants is still emerging [65]. Due to the effectiveness of CRISPR/Cas technology, it can be also used as an influential genome-editing tool for genetic modification and functional characterization of plant genes/miRNAs and for genome modification to improve agricultural crops [66].

7. Conclusion and future perspectives

Above mentions exemplified case studies indicated that currently miRNAs viewed as a most essential gene regulator tool. This chapter also focused on the studies that describe the multipurpose role of miRNAs in plants. Recent advancement in biological science made to access and characterize miRNAs in crop plants, with a growing number of researches on the significant function of miRNAs [49]. As described, in plants few important processes like homeostasis, growth and development, vegetative to reproductive phase transition and signal transduction and response to various stress are regulated by miRNA [13]. Few recent studies also showed that in plants miRNAs biomolecule act as plant defense and organizers of immune responses [67]. In this regard, the supervision of miRNA expression levels would recognize a crucial way for enhancing the plant growth and development as well as various biotic and abiotic stresses tolerance.

Various transgenic approaches, focused on miRNA role and its importance and identification on corresponding target genes. These includes miRNA overexpression, tissue-specific expression of the targets or miRNAs or stress induce, artificial miRNA, expression of miRNA resistant target gene and artificial target mimic [68]. In cases where the natural target gene has a harmful effect on plant, the overexpression of the regulatory miRNA decreasing the corresponding mRNA expression [47]. Where the target gene has a positive effect, approaches tracked can involve the artificial target mimics or selection of miRNA resistant target gene and the overexpression of the target genes [37]. Although this approach generally successful, the applied agricultural application of the miRNA methods is interesting since the modification and alteration in complicated multi-genic traits such as productivity may require alteration in expression of different genes during different developmental stages of plant. However, for improving crop stress tolerance several new strategies used for improving miRNA bases gene regulation in model plants [69].

It is also possible that few miRNAs affect expression of target gene only in some specific cell, and exclusively under specific environmental conditions as several miRNAs may have complicated expression pattern. In this circumstance, analysis of effect of miRNA on expression in whole organs would uncertain; therefore, the experiment should be specifically and carefully designed for improved results. Several artificial miRNAs designed and used to suppress expression of a target gene and protein-coding mRNA of interest is one of the among valuable and suitable approach for plant improvement [70]. Due to insufficient concentration and expression the effect of some crucial miRNAs may not visible in living tissue. For producing plants with desirable characters, it would be essential to execute some quantitative analysis of the natural miRNA(s) of the cells, before designing the artificial miRNAs.

Although, old artificial miRNA approach used as important tool for genetic modification [70], but still there is some more knowledge required for its appropriate use in the differentiation and translational inhibition by miRNA. Single miRNA produced by artificial nucleic acid with the ability to inhibit target loci, avoid and predict all the unnecessary genes for experiment design [52]. By modifying miRNA regulatory pathways, miRNA activity also altered for attaining a desired trait [71]. miRNA action negatively controlled by endogenous RNAs, it is a more flexible and fast way to understand the miRNA function, as well as for manipulation of target gene in plants [72]. This technique offers the way to attenuated miRNA inactivation by significant regulation of native miRNA targets and produce a wide range of phenotypes by change the miRNA decoy site sequence. However, for appropriate crop improvement and proper utilization of this approach, miRNA inactivation at appropriate level and to avoid off-site effects of miRNA which generate some false positive results [73]. Also, the interaction between miRNA and miRNA decoy, as well as the several complex produce by this interaction, is not well known and can causes the miRNA destabilization in organisms. Consequently, for the practical uses of this method in plants, some new and advance modification is needed on the procedure basic decoy-associated regulation of gene expression and miRNA turnover. miRNA and miRNA variants act as important tool in plant improvement, it is also nullifying the side effect produced accidently by genetic engineering technology [73]. If the targeted gene and expression of miRNA is modified, it can produce some pleiotropic alteration in plant development and morphogenesis. Therefore, it is very essential to recognize the miRNA regulation method for plant development, growth and plant responses to different biotic and abiotic stresses. This will simplify the plan of appropriate strategies resulting in the desired traits but with minimum trade-offs in the modified crops.

Author details

Jyoti Rani
Department of Botany, Government College for Women, Bhiwani, Haryana, India

*Address all correspondence to: jyoti.sangwan@yahoo.com

References

[1] Sanan-Mishra, N., Kumar, V., Sopory, S. K., and Mukherjee, S. K. (2009). Cloning and validation of novel miRNA from basmati rice indicates cross talk between abiotic and biotic stresses. Mol. Genet. Genom. 282, 463-474.

[2] Djami-Tchatchou, A. T., and Dubery, I. A. (2015). Lipopolysaccharide perception leads to dynamic alterations in the microtranscriptomes of *Arabidopsis thaliana* cells and leaf tissues. BMC Plant Biol. 15, 79.

[3] Navarro, L., Dunoyer, P., Jay, F., Arnold, B., Dharmasiri, N., Estelle, M., et al. (2006). A plant miRNA contributes to antibacterial resistance by repressing auxin signaling. Science. 312, 436-439.

[4] Shriram, V., Kumar, V., Devarumath, R. M., Khare, T. S., and Wani, S. H. (2016). MicroRNAs as potential targets for abiotic stress tolerance in plants. Front. Plant Sci. 7: 8-17.

[5] Ku, Y. S., Wong, J. W. H., Mui, Z., Liu, X., Hui, J. H. L., Chan, T. F., et al. (2015). Small RNAs in plant responses to abiotic stresses: regulatory roles and study methods. Int. J. Mol. Sci. 16, 24532-24554.

[6] Jones-Rhoades, M. W., Bartel, D. P., and Bartel, B. (2006). MicroRNAs and their regulatory roles in plants. Ann. Rev. Plant Biol. 57, 19-53.

[7] Pantaleo, V., Szittya, G., Moxon, S., Miozzi, L., Moulton, V., Dalmay, T., et al. (2010). Identification of grapevine microRNAs and their targets using high-throughput sequencing and degradome analysis. Plant J. 62, 960-976.

[8] Sun, G., Stewart, C. N. Jr., Xiao, P., and Zhang, B. (2012). MicroRNA expression analysis in the cellulosic biofuel crop Switchgrass (*Panicum virgatum*) under abiotic stress. PLoS one. 7, 1-7.

[9] Pelaez, P., Trejo, M. S., Iñiguez, L. P., Estrada-Navarrete, G., Covarrubias, A. A., Reyes, J. L., et al. (2012). Identification and characterization of microRNAs in *Phaseolus vulgaris* by high-throughput sequencing. BMC Genomic. 13, 1-8.

[10] Arribas-Hernández, L., Marchais, A., Poulsen, C., Haase, B., Hauptmann, J., Benes, V., et al. (2016). The slicer activity of ARGONAUTE1 is required specifically for the phasing, not production, of trans-acting short interfering RNAs in *Arabidopsis*. Plant Cell. 28, 1563-1580.

[11] Iki, T. (2017). Messages on small RNA duplexes in plants. J. Plant Res. 130, 7-16.

[12] Cuperus, J. T., Fahlgren, N., and Carrington, J. C. (2011). Evolution and functional diversification of miRNA genes. Plant Cell. 23, 431-442.

[13] Sun, G. (2012). MicroRNAs and their diverse functions in plants. Plant Mol. Biol. 80, 17-36.

[14] Fahlgren, N., Howell, M. D., Kasschau, K. D., Chapman, E. J., Sullivan, C. M., Cumbie, J. S., et al. (2007). High-throughput sequencing of *Arabidopsis* microRNAs: evidence for frequent birth and death of miRNA genes. PLoS one. 2, 1-14.

[15] Lelandais-Brière, C., Sorin, C., Declerck, M., Benslimane, A., Crespi, M., and Hartmann, C. (2010). Small RNA diversity in plants and its impact in development. Curr. Genomics. 11, 14-23.

[16] Guleria, P., Mahajan, M., Bhardwaj, J., and Yadav, S. (2011). Plant Small RNAs: biogenesis, mode of action and

their roles in abiotic stresses. Genom. Proteom. Bioinform. 9, 183-199.

[17] Jeong, D. H. (2016). Functional diversity of microRNA variants in plants. J. Plant Biol. 59, 303-310.

[18] Starega-Roslin, J., Krol, J., Koscianska, E., Kozlowski, P., Szlachcic, W. J., Sobczak, K., et al. (2011). Structural basis on microRNA length variety. Nucl. Acids Res. 39, 257-268.

[19] Schommer, C., Bresso, E. G., Spinelli, S. V., and Palatnik, J. F. (2012). "Role of microRNA miR319 in plant development. "MicroRNAs in Plant Development and Stress Responses"," in Signaling and Communication in Plants. Berlin: Springer-Verlag. 15, 29-47.

[20] Reinhart, B. J., Weinstein, E. G., Rhoades, M. W., Bartel, B., and Bartel, D. P. (2002). MicroRNAs in plants. Genes Dev. 16, 1616-1626.

[21] Aukerman, M. J., and Sakai, H. (2003). Regulation of flowering time and floral organ identity by a microRNA and its APETALA2-like target genes. Plant Cell. 15, 2730-2741.

[22] Palatnik, J. F., Allen, E., Wu, X., Schommer, C., Schwab, R., Carrington, J. C., et al. (2003). Control of leaf morphogenesis by microRNAs. Nature. 425, 257-263.

[23] Guo, X., Gui, Y., Wang, Y., Zhu, Q. H., Helliwell, C., and Fan, L. (2008). Selection and mutation on microRNA target sequences during rice evolution. BMC Genomics, 6, 1-15.

[24] Han, J., Kong, M. L., Xie, H., Sun, Q. P., Nan, Z. J., Zhang, Q. Z., et al. (2013). Identification of miRNAs and their targets in wheat (*Triticum aestivum* L.) by EST analysis. Genet. Mol. Res. 12, 3793-3805.

[25] Zhang, B. H., Wang, Q. L., Wang, K. B., Pan, X. P., Liu, F., Guo, T. L., et al. (2007). Identification of cotton microRNAs and their targets. Gene. 397, 26-37.

[26] Xie, F., Frazier, T. P., and Zhang, B. (2011). Identification, characterization and expression analysis of MicroRNAs and their targets in the potato (*Solanum tuberosum*). Gene. 473, 8-22.

[27] Matts, J., Jagadeeswaran, G., Roe, B. A., and Sunkar, R. (2010). Identification of microRNAs and their targets in switchgrass, a model biofuel plant species. J. Plant Physiol. 167, 896-904.

[28] Kulcheski, F. R., de Oliveira, L. F. V., Molina, L. G., Almerão, M. P., Rodrigues, F. A., Marcolino, J., et al. (2011). Identification of novel soybean microRNAs involved in abiotic and biotic stresses. BMC Genomics. 12, 307.

[29] Wang, M., Wang, Q., and Wang, B. (2012). Identification and characterization of microRNAs in Asiatic cotton (*Gossypium arboretum* L.). PLoS one. 7, 1-8.

[30] Zhang, B., and Wang, Q. (2015). MicroRNA-based biotechnology for plant improvement. J. Cell. Physiol. 230, 1-15.

[31] Zhang, Z., Yu, J., Li, D., Zhang, Z., Liu, F., Zhou, X., et al. (2010). PMRD: plant microRNA database. Nucleic Acids Res. 38, D806–D813.

[32] Steffens, N. O., Galuschka, C., Schindler, M., Bülow, L., and Hehl, R. (2004). AthaMap: an online resource for in silico transcription factor binding sites in the *Arabidopsis thaliana* genome. Nucleic Acids Res. 32, 368-372.

[33] Zhang, Y., Wang, W., Chen, J., Liu, J., Xia, M., and Shen, F. (2015). Identification of miRNAs and their targets in cotton inoculated with *Verticillium dahliae* by high-throughput sequencing and degradome analysis. Int. J. Mol. Sci. 16, 14749-14768.

[34] Shriram, V., Kumar, V., Devarumath, R. M., Khare, T. S., and Wani, S. H. (2016). MicroRNAs as potential targets for abiotic stress tolerance in plants. Front. Plant Sci. 7: 8-17.

[35] Xie, F., Stewart, C. N. Jr., Taki, F. A., He, Q., Liu, H., and Zhang, B. (2014). High-throughput deep sequencing shows that microRNAs play important roles in switchgrass responses to drought and salinity stress. Plant Biotechnol. J. 12, 354-366.

[36] Mallory, A. C., and Vaucheret, H. (2006). Functions of microRNAs and related small RNAs in plants. Nat. Genet. 38, S31–S36.

[37] Gupta, P. K. (2015). MicroRNAs and target mimics for crop improvement. Curr. Sci. 108, 1624-1633.

[38] Sanan-Mishra, N., and Mukherjee, S. K. (2007). A peep into the plant miRNA world. Open Plant Sci. J. 1, 1-9.

[39] Franco-Zorrilla, J. M., Valli, A., Todesco, M., Mateos, I., Puga, M. I., Rubio Somoza, I., et al. (2007). Target mimicry provides a new mechanism for regulation of microRNA activity. Nat. Genet. 39, 1033-1037.

[40] Khraiwesh, B., Ossowski, S., Weigel, D., Reski, R., and Frank, W. (2008). Specific gene silencing by artificial microRNAs in *Physcomitrella patens*: an alternative to targeted gene knockouts. Plant Physiol. 148, 684-693.

[41] Qu, J., Ye, J., and Fang, R. X. (2007). Artificial microRNA-mediated virus resistance in plants. J. Virol. 81, 6690-6699.

[42] Sharma, N., Tripathi, A., and Sanan-Mishra, N. (2015). Profiling the expression domains of a rice-specific microRNA under stress. Front. Plant Sci. 6, 333.

[43] Bej, S., and Basak, J. (2014). MicroRNAs: the potential biomarkers in plant stress response. Am. J. Plant Sci. 5, 748-759.

[44] Zhang, X., Zou, Z., Gong, P. J., Zhang, J. H., Ziaf, K., Li, H. X., et al. (2011). Overexpression of microRNA169 confers enhanced drought tolerance to tomato. Biotechnol. Lett. 33, 403-409.

[45] Gao, P., Bai, X., Yang, L., Lv, D. K., Li, Y., Cai, H., et al. (2010). Over-expression of osa-MIR396c decreases salt and alkali stress tolerance. Planta 231, 991-1001.

[46] Lu, Y. Z., Feng, Z., Bian, L. Y., Xie, H., and Liang, J. S. (2011). miR398 regulation in rice of the responses to abiotic and biotic stresses depends on CSD1 and CSD2 expression. Funct. Plant Biol. 38, 44-53.

[47] Yang, C., Li, D., Mao, D., Liu, X., Ji, C., Li, X., et al. (2013). Overexpression of microRNA319 impacts leaf morphogenesis and leads to enhanced cold tolerance in rice (*Oryza sativa* L.). Plant Cell Environ. 36, 2207-2218.

[48] Campo, S., Peris-Peris, C., Sire, C., Moreno, A. B., Donaire, L., Zytnicki, M., et al. (2013). Identification of a novel microRNA (miRNA) from rice that targets an alternatively spliced transcript of the Nramp6 (Natural resistance-associated macrophage protein 6) gene involved in pathogen resistance. New Phytol. 199, 212-227.

[49] Li, Y., Zhang, Q. Q., Zhang, J., Wu, L., Qi, Y., and Zhou, J. M. (2010). Identification of microRNAs involved in pathogen-associated molecular pattern-triggered plant innate immunity. Plant Physiol. 152, 2222-2231.

[50] Djami-Tchatchou, A. T., Mishra, N.S., Ntushelo, K., and Dubery, I. A. (2017). Functional role of miRNAs in agronomically important

plants-potential as targets for crop improvement and protection. Front. Plant Sci. 8, 1-24.

[51] Ori, N., Cohen, A. R., Etzioni, A., Brand, A., Yanai, O., Shleizer, S., et al. (2007). Regulation of LANCEOLATE by miR319 is required for compound-leaf development in tomato. Nature Genet. 39, 787-791.

[52] Schwab, R., Palatnik, J. F., Riester, M., Schommer, C., Schmid, M., and Weigel, D. (2005). Specific effects of microRNAs on the plant transcriptome. Dev. Cell 8, 517-527.

[53] Fu, C., Sunkar, R., Zhou, C., Shen, H., Zhang, J. Y., Matts, J., et al. (2012). Overexpression of miR156 in switchgrass (*Panicum virgatum* L.) results in various morphological alterations and leads to improved biomass production. Plant Biotechnol. J. 10, 443-452.

[54] Xie, K. B., Shen, J. Q., Hou, X., Yao, J. L., Li, X. H., Xiao, J. H., et al. (2012). Gradual increase of miR156 regulates temporal expression changes of numerous genes during leaf development in rice. PlantPhysiol.158,1382-1394.

[55] Zhang, Y. C., Yu, Y., Wang, C. Y., Li, Z. Y., Liu, Q., Xu, J., et al. (2013). Overexpression of microRNA OsmiR397 improves rice yield by increasing grain size and promoting panicle branching. Nat. Biotechnol. 31, 848-852.

[56] Yoon, E. K., Yang, J. H., Lim, J., Kim, S. H., Kim, S. K., and Lee, W. S. (2010). Auxin regulation of the microRNA390-dependent transacting small interfering RNA pathway in *Arabidopsis* lateral root development. Nucleic Acids Res. 38, 1382-1391.

[57] Bustos-Sanmamed, P., Mao, G. H., Deng, Y., Elouet, M., Khan, G. A., Bazin, J., et al. (2013). Overexpression of miR160 affects root growth and nitrogen fixing nodule number in *Medicago truncatula*. Funct. Plant Biol. 40, 1208-1220.

[58] Lauter, N., Kampani, A., Carlson, S., Goebel, M., and Moose, S. P. (2005). MicroRNA172 down-regulates glossy15 to promote vegetative phase change in maize. Proc. Natl. Acad. Sci. U.S.A. 102, 9412-9417.

[59] Chuck, G., Cigan, A. M., Saeteurn, K., and Hake, S. (2007). The heterochronic maize mutant Corngrass1 results from overexpression of a tandem microRNA. Nature Genet. 39, 544-549.

[60] Kumar, V., Shriram, V., Kavi Kishor, P. B., Jawali, N., and Shitole, M. G. (2010). Enhanced proline accumulation and salt stress tolerance of transgenic indica rice by over-expressing P5CSF129A gene. Plant Biotechnol. Rep. 4, 37-48.

[61] Khan, Z., Khan, S. H., Mubarik, M. S., Sadia, B., and Ahmad, A. (2017). Use of TALEs and TALEN technology for genetic improvement of plants. Plant Mol. Biol. Rep. 35, 1-19.

[62] Chan, H., Yi, B., Ma, R., Zhang, X., Zhao, H., and Xi, Y. (2016). CRISPR/cas9, a novel genomic tool to knock down microRNA in vitro and in vivo. Sci. Rep. 6, 1-12.

[63] Xie, K., and Yang, Y. (2013). RNA-guided genome editing in plants using a CRISPR–Cas system. Mol. Plant. 6, 1975-1983.

[64] Jacobs, T. B., La Fayette, P. R., Schmitz, R. J., and Parrott, W. A. (2015). Targeted genome modifications in soybean with CRISPR/Cas9. BMC Biotechnol. 15: 16.

[65] Schiml, S., and Puchta, H. (2016). Revolutionizing plant biology: multiple ways of genome engineering by CRISPR/Cas. Plant Methods. 12, 1-8.

[66] Bassak, J., and Nithin, C. (2015). Targeting non-coding RNAs in plants with the CRISPR-Cas technology is a challenge yet worth accepting. Front. Plant Sci. 6:1001.

[67] Balmer, D., and Mauch-Mani, B. (2013). Small yet mighty – microRNAs in plant microbe interactions. Microrna. 2, 72-79.

[68] Zhou, M., and Luo, H. (2013). MicroRNA-mediated gene regulation: potential applications for plant genetic engineering. Plant Mol. Biol. 83, 59-75.

[69] Song, J. B., Gao, S., Wang, Y., Li, B. W., Zhang, Y. L., and Yang, Z. M. (2016). miR394 and its target gene LCR are involved in cold stress response in Arabidopsis. Plant Gene. 5, 56-64.

[70] Khraiwesh, B., Ossowski, S., Weigel, D., Reski, R., and Frank, W. (2008). Specific gene silencing by artificial microRNAs in *Physcomitrella patens*: an alternative to targeted gene knockouts. Plant Physiol. 148, 684-693.

[71] Ivashuta, S., Banks, I. R., Wiggins, B. E., Zhang, Y., Ziegler, T. E., Roberts, J. K., et al. (2011). Regulation of gene expression in plants through miRNA inactivation. PLoS one. 6, 1-11.

[72] Banks, I. R., Zhang, Y., Wiggins, B. E., Heck, G. R., and Ivashuta, S. (2012). RNA decoys: an emerging component of plant regulatory networks? Plant Signal. Behav. 7, 1188-1193.

[73] Kamthan, A., Chaudhuri, A., Kamthan, M., and Datta, A. (2015). Small RNAs in plants: recent development and application for crop improvement. Front. Plant Sci. 6, 1-17.